Michael Valentine Ball

Essentials of Bacteriology

Being a concise and systematic Introduction to the Study of Micro-Organisms.

Second Edition

Michael Valentine Ball

Essentials of Bacteriology
Being a concise and systematic Introduction to the Study of Micro-Organisms. Second Edition

ISBN/EAN: 9783337176976

Printed in Europe, USA, Canada, Australia, Japan

Cover: Foto ©berggeist007 / pixelio.de

More available books at **www.hansebooks.com**

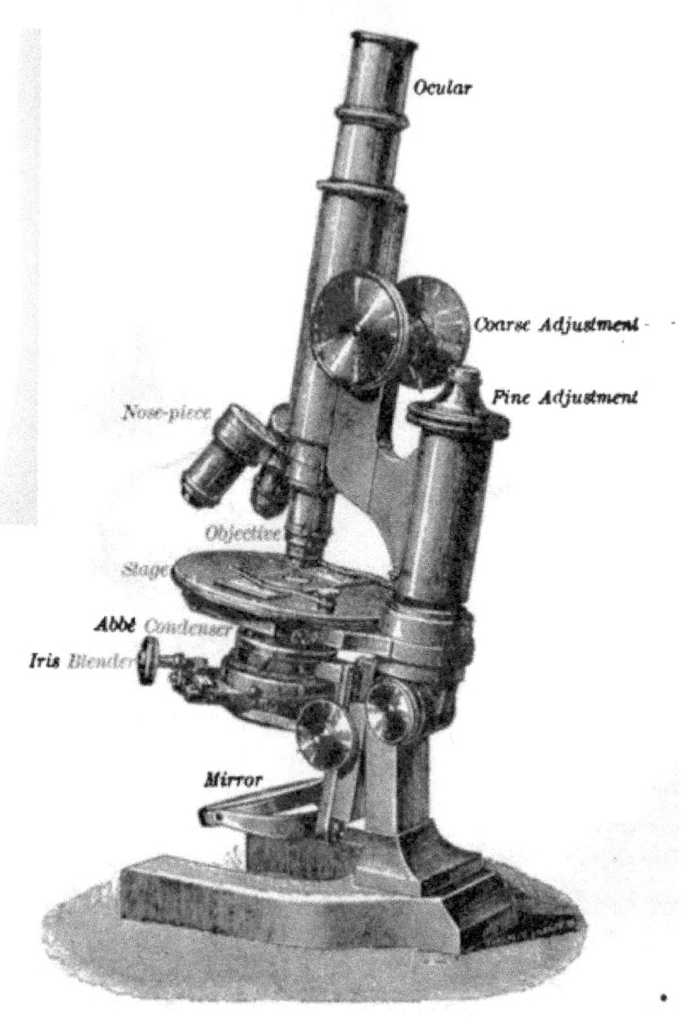

BACTERIOLOGICAL MICROSCOPE (WITH ABBÉ AND BLENDER IN POSITION).

ESSENTIALS OF BACTERIOLOGY:

BEING A

CONCISE AND SYSTEMATIC INTRODUCTION TO THE STUDY OF MICRO-ORGANISMS

FOR THE USE OF

STUDENTS AND PRACTITIONERS.

BY

M. V. BALL, M.D.,

PHYSICIAN TO THE EASTERN STATE PENITENTIARY AT PHILADELPHIA.

SECOND EDITION.

WITH EIGHTY-ONE ILLUSTRATIONS, SOME IN COLORS, AND FIVE PLATES.

LONDON:
HENRY KIMPTON
82, HIGH HOLBORN, W. C.
1895.
(Printed in U. S. A.)

PREFACE TO SECOND EDITION.

In this second edition, the results of last year's earnest work have been embodied. The efforts of Bacteriologists, since Koch's tuberculin announcement, have been directed to the elaboration of therapeutic agents from the chemical products of Bacteria. Physiological chemistry can accomplish more here than Bacteriology. The separation from the blood of Antitoxines, and their application to the cure of disease, will probably revolutionize our present method of treatment, and add some peculiar agents to our Materia Medica. The question of immunity is still unanswered, though the Phagocytic theory of Metschnikoff and the Alexines of Buchner are bringing us to the solution. Upon these lines the greatest efforts are at present being made.

We are very grateful to all those who have taken notice of our little effort, and we hope to merit a continuance of their regard.

The way in which the first edition was received assured us of its success.

M. V. B.

PREFACE TO FIRST EDITION.

FEELING the need of a Compendium on the subject of this work, it has been our aim to produce a concise treatise upon the Practical Bacteriology of *to-day*, chiefly for the medical student, which he may use in his laboratory.

It is the result of experience gained in the Laboratory of the Hygienical Institute, Berlin, under the guidance of Koch and Fränkel; and of information gathered from the original works of other German, as well as of French, bacteriologists.

Theory and obsolete methods have been slightly touched upon. The scope of the work, and want of space, forbade adequate consideration of them. The exact measurements of bacteria have not been given. The same bacterium varies often much in size, owing to differences in the media, staining, etc.

We have received special help from the following books, which we recommend to students for further reference:—

MACÉ: Traité pratique de Bacteriologie.
FRÄNKEL: Grundriss der Bakterienkunde.
EISENBERG: Bakteriologische Diagnostik.
CROOKSCHANK, E. M.: Manual of Bacteriology.
GUNTHER: Einführing in das Studium der Bacteriologie, etc.
WOODHEAD AND HARE: Pathological Mycology.
SALMONSEN: Bacteriological Technique (English translation).

M. V. BALL.

BUFFALO, N. Y., October 1, 1891.
62 Delaware Avenue.

CONTENTS.

PART I.

GENERAL CONSIDERATIONS AND TECHNIQUE.

PART II.

SPECIAL BACTERIOLOGY.

APPENDIX.

INTRODUCTION.

History.—The microscope was invented about the latter part of the sixteenth century; and soon after, by its aid, minute organisms were found in decomposing substances. Kircher, in 1646, suggested that diseases might be due to similar organisms; but the means at his disposal were insufficient to enable him to prove his theories. Anthony Van Leuwenhoeck, of Delft, Holland (1680 to 1723), so improved the instrument that he was enabled thereby tc discover micro-organisms in vegetable infusion, saliva, fecal matter, and scrapings from the teeth. He distinguished several varieties, showed them to have the power of locomotion, and compared them in size with various grains of definite measurement. It was a great service that this "Dutch naturalist" rendered the world; and he can rightly be called the "father of microscopy."

Various theories were then formulated by physicians to connect the origin of different diseases with bacteria; but no proofs of the connection could be obtained. Andry, in 1701, called bacteria *worms*. Müller, of Copenhagen, in 1786, made a classification composed of two main divisions—monas and vibrio; and with the aid of the compound microscope was better able to describe them. Ehrenberg, in 1833, with still better instruments, divided bacteria into four orders: bacterium, vibrio, spirillum, and spirochæte. It was not until 1863 that any positive advance was made in connecting bacteria with disease. Rayer and Davaine had in 1850 already found a rod-shaped bacterium in the blood of animals suffering from *splenic fever* (*sang de rate*), but they attached no special significance to their discovery until Pasteur made public his grand researches in regard to fermentation and the role bacteria played in the economy. Then Davaine resumed his studies, and in 1863 established by experiments the bacterial nature of splenic fever or anthrax.

But the first complete study of a contagious affection was made by Pasteur in 1869, in the diseases affecting silk-worms—pebrine and flacherie—which he showed to be due to micro-organisms.

Then Koch, in 1875, described more fully the anthrax bacillus, gave a description of its spores and the properties of the same, and was enabled to cultivate the germ on artificial media; and, to complete the chain of evidence, Pasteur and his pupils supplied the last link by reproducing the same disease in animals by artificial inoculation from pure cultures. The study of the bacterial nature of anthrax has been the basis of our knowledge of all contagious maladies, and most advances have been made first with the bacterium of that disease.

Since then bacteriology has grown to huge proportions—become a science of itself—and thousands of earnest workers are adding yearly solid blocks of fact to the structure, which structure it will be our aim to briefly describe in the pages which are to follow.

ESSENTIALS OF BACTERIOLOGY.

PART I.

GENERAL CONSIDERATIONS.

CHAPTER I.

BACTERIA.

BACTERIA (βακτηριον, little staff) is the name given to a group of the lowest form of plants, very closely following the algæ. They were called *Fission-Fungi* or *Schizomycetes* (σχιζω, to cleave, μυκης, fungus), because it was thought that, as the Fungi, they lived without the chlorophyll. The word fission was supplied to distinguish them from moulds and yeasts, it denoting the manner of reproduction. Since several bacteria have been found to possess chlorophyll, and as a great many increase in other ways than by simple fission—the name of Schizomycetes can no longer be applied, though the word Bacteria leaves much to be desired.

Classification. Ferdinand Cohn, in the middle of the present century, was the first to demonstrate bacteria to be of vegetable origin, they being placed previous to that among the infusoria. He arranged them according to their form under four divisions.

Cohn's System. I. Spherobacteria (globules).
II. Microbacteria (short rods).
III. Desmobacteria (long rods).
IV. Spirobacteria (spirals).

As expressed at the present time, Micrococcus, Bacillus, and Spirillum. This classification is very superficial, but because a better one has not been found it is most in use to-day.

De Bary's System. De Bary divides bacteria into two groups, those arising from or giving rise to endospores and those developed from arthrospores. This division has a more scientific value than the first.

Fig. 1.

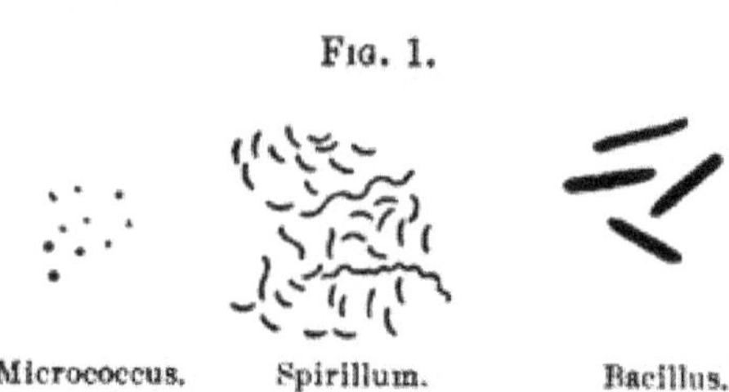

Micrococcus. Spirillum. Bacillus.

Structure. Bacteria are cells; they appear as round or cylindrical of an average diameter or transverse section of 0.001 mm. (=1 micromillimeter), written 1 μ. The cell, as other plant-cells, is composed of a membranous cell-wall and cell-contents; "cell-nuclei" have not yet been observed, but the latest researches point to their presence.

Cell-Wall. The cell-wall is composed of plant cellulose, which can be demonstrated in some cases by the tests for cellulose. The membrane is firm and can be brought plainly into view by the action of iodine upon the cell-contents which contracts them.

Cell-Contents. The contents of the cell consist mainly of protoplasm, usually homogeneous, but in some varieties, finely granular, or holding pigment, chlorophyll, granulose, and sulphur in its structure.

It is composed chiefly of *mycoprotein.*

Gelatinous Membrane. The outer layer of the cell-membrane can absorb water and become gelatinoid, forming either a little envelope or capsule around the bacterium or preventing the separation of the newly-branched germs, forming chains and bunches, as *strepto- and staphylo-cocci.* Long filaments are also formed.

Zooglœa. When this gelatinous membrane is very thick, irregular masses of bacteria will be formed, the whole growth being in one jelly-like lump. This is termed a zooglœa (ζωον, animal, γλοιος, glue).

Locomotion. Many bacteria possess the faculty of self-movement, carrying themselves in all manner of ways across the microscopic field, some very quickly, others leisurely.

Vibratory Movements. Some bacteria vibrate in themselves, appearing to move, but they do not change their place; these movements are denoted as molecular or "*Brownian.*"

FIG. 2.

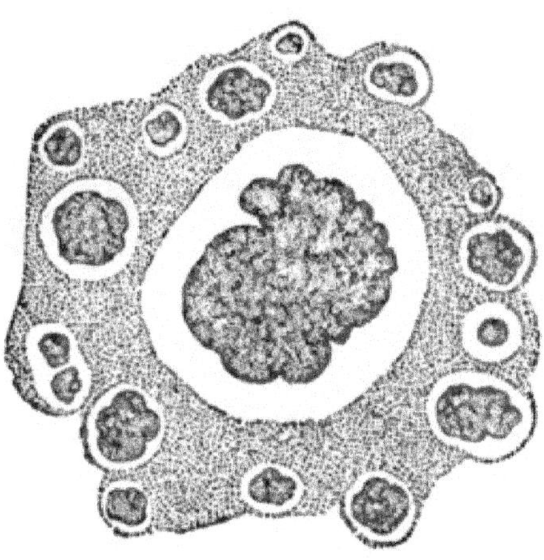

Zoogloea.

Flagella. Little threads or lashes are found attached to many of the motile bacteria, either at the poles or along the sides, sometimes only one, and on some several, forming a tuft.

These flagella are in constant motion and can probably be considered as the organs of locomotion; they have not yet been discovered upon all the motile bacteria, owing no doubt to our imperfect methods of observation. They can be stained and have been photographed. See Fig. 3.

Reproduction. Bacteria multiply either through simple *division* or through *fructification* by means of small round or oval bodies called spores from *spora* (seed.) In the first case, *division*, the cell elongates, and at one portion, usually the middle, the cell-wall indents itself gradually, forming a septum and dividing the cell into two equal parts, just as occurs in the higher plant and animal cells. See Fig. 4.

Fig. 3.

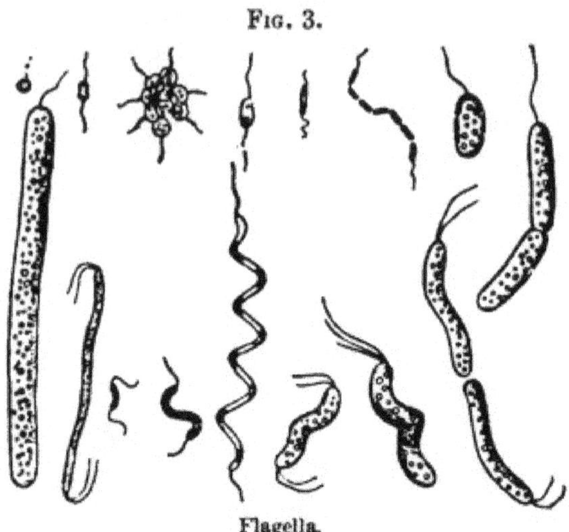

Flagella.

Successive divisions take place, the new members either existing as separate cells or forming part of a community or group.

Fig. 4.

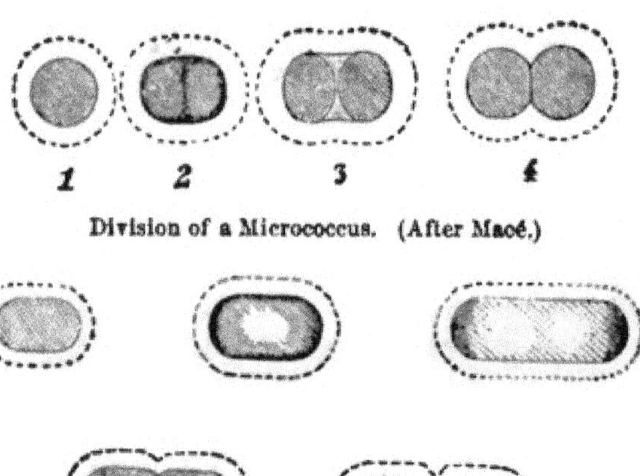

Division of a Micrococcus. (After Macé.)

Division of a Bacillus. (After Macé.)

Spore Formations. Two forms of sporulation, *Endosporous* and *Arthrosporous.* First, a small granule develops in the protoplasm of a bacterium, this increases in size, or several little granules coalesce to form an elongated, highly refractive, clearly defined object, rapidly attaining its real size, and this is the spore. The remainder of the cell-contents has now disappeared, leaving the spore in a dark, very resistant, membrane or capsule, and beyond this the weak cell-wall. The cell-wall dissolves gradually or stretches and allows the spore to be set free.

Each bacterium gives rise to but one spore. It may be at either end or in the middle (Fig. 5). Some rods take on a peculiar shape at the site of the spore, making the rod look like a drum-stick or spindle, clostridium (Fig. 6).

Fig. 5. Fig. 6.

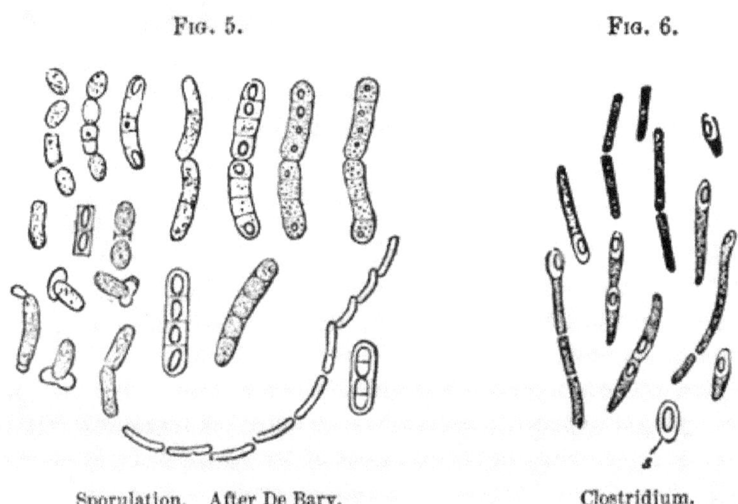

Sporulation. After De Bary. Clostridium.

Spore Contents. What the real contents of spores are is not known. In the mother cell at the site of the spore little granules have been found which stain differently from the rest of the cell, and these are supposed to be the beginnings, the *sporogenic bodies.* The most important part of the spore is its *capsule;* to this it owes its resisting properties. It consists of two separate layers, a thin membrane around the cell, and a firm outer gelatinous envelope.

Germination. When brought into favorable conditions, the spore begins to lose its shining appearance, the outer firm membrane begins to swell, and it now assumes the shape and size of the cell from which it sprang, the capsule having burst, so as to allow the young bacillus to be set free.

Requisites for Spore Formation. It was formerly thought that when the substratum could no longer maintain it, or had become infiltrated with detrimental products, the bacterium-cell produced spores, or rather turned itself into a spore to escape annihilation; but we know now that only when the conditions are the most favorable to the well-being of the cell, does it produce fruit, just as with every other type of plant or animal life, a certain amount of oxygen and heat being necessary for good spore formation.

Asporogenic Bacteria. Bacteria can be so damaged that they will remain sterile, not produce any spores. This condition can be temporary only, or permanent.

Arthrosporous. All the above remarks relate to Endospores, spores that arise within the cells.

In the other group called Arthrospores, individual members of a colony or aggregation leave the same, and become the originators of new colonies, thus assuming the character of spores.

The Micrococci furnish examples of this form.

Some authorities have denied the existence of the arthrosporous formation.

Resistance of Spores. Because of the very tenacious envelope, the spore is not easily influenced by external measures. It is said to be the most resisting object of the organic world.

Chemical and physical agents that easily destroy other life have very little effect upon it.

Many spores require a temperature of 140° C. dry heat for several hours to destroy them. The spores of a variety of potato-bacillus (bacillus mesentericus) can withstand the application of steam at 100° C. for four hours.

CHAPTER II.

ORIGIN OF BACTERIA AND THEIR DISTRIBUTION.

As Pasteur has shown, all bacteria develop from pre-existing bacteria, or the spores of the same. They cannot, do not arise *de novo.*

Their wide and almost universal diffusion is due to the minuteness of the cells and the few requirements for their existence.

Very few places are free from germs; the air on the high seas, and on the mountain tops, is said to be free from bacteria, but it is questionable.

One kind of bacterium will not produce another kind.

A bacillus does not arise from a micrococcus or the typhoid fever bacillus produce the bacillus of tetanus.

This subject has been long and well discussed, and it would take many pages to state the "pros" and "cons," therefore, this positive statement is made, it being the position now held by the principal authorities.

Saprophytes and Parasites. (*Saprophytes*, ςαπρος, putrid, φυτον, plant. *Parasites*, παρα, aside of διτος, food.) Those bacteria which live on the dead remains of organic life are known as Saprophytic Bacteria, and those which choose the living bodies of their fellow-creatures for their habitat are called Parasitic Bacteria. Some, however, develop equally well as Saprophytes and Parasites. They are called *Facultative Parasites.*

Conditions of Life and Growth of Bacteria. *Influence of Temperature.*—In general, a temperature ranging from 10° C. to 40° C. is necessary to their life and growth.

Saprophytes take the lower temperatures; Parasites, the temperature more approaching the animal heat of the warm-blooded. Some forms require a nearly constant heat, growing within very small limits, as the Bacillus of Tuberculosis.

Some forms can be arrested in their development by a warmer or colder temperature, and then restored to activity by a return to the natural heat.

A few varieties exist only at freezing point of water; and others again that will not live under a temperature of 60° C.

For the majority of Bacteria a temperature of 60° C. is destructive; and several times freezing and thawing very fatal.

Influence of Oxygen.—Two varieties of bacteria in relation to oxygen.

The one *ærobic*, growing in air; the other *anærobic*, living without air.

Obligate ærobins, those which exist only when oxygen is present.

Facultative ærobins, those that live best when oxygen is present, but can live without it.

Obligate or *true anærobins*, those which cannot exist where oxygen is.

Facultative anærobins, those which exist better where there is no oxygen, but can live in its presence.

Some derive the oxygen which they require out of their nutriment, so that a bacterium may be ærobic and yet not require the presence of free oxygen.

Ærobins may consume the free oxygen of a region and thus allow the anærobins to develop. By improved methods of culture many varieties of anærobins have been discovered.

Influence of Light.—Sunlight is very destructive to bacteria. A few hours' exposure to the sun has been fatal to anthrax bacilli, and the cultures of bacillus tuberculosis have been killed by a few days' standing in daylight.

Effects of Electricity.—Electricity arrests growth.

Vital Actions of Microbes. Bacteria feeding upon organic compounds produce chemical changes in them, not only by the withdrawal of certain elements, but also by the excretion of these elements changed by digestion. Sometimes such changes are destructive to themselves, as when lactic and butyric acids are formed in the media.

Oxidation and reduction are carried on by some bacteria. Ammonia, hydrogen sulphide, and trimethylamin are a few of the chemical products produced by bacteria.

Ptomaines. Brieger found a number of complex alkaloids, closely resembling those found in ordinary plants, and which he named ptomaïnes, from πτῶμα (corpse), because obtained from putrefying objects.

Fermentation. This form of "splitting up"—fermentation, as it is called—is due to the direct action of vegetable organisms. Many bacteria have the power of ferments.

Putrefaction. When fermentation is accompanied by development of offensive gases, a decomposition occurs, which is called putrefaction, and this, in organic substances, is due entirely to bacteria.

Liquefaction of Solid Gelatine. Some varieties of bacteria digest the nutrient gelatine, and so dissolve it; others excrete a ferment which liquefies the gelatine.

Producers of Disease. Various pathological processes are caused by bacteria, the name given to such diseases being *infectious diseases* and the germs themselves called disease-producing *pathogenic bacteria.* Those which do not form any pathological process are called *non-pathogenic bacteria.*

Pigmentation. Some bacteria are endowed with the property of forming pigments either in themselves, or producing a chromogenic body which, when set free, gives rise to the pigment. In some cases the pigments have been isolated and many of the properties of the aniline dyes discovered in them.

Phosphorescence. Many bacteria have the power to form light, giving to various objects which they inhabit a characteristic glow or phosphorescence.

Fluorescence. An iridescence, or play of colors, develops in some of the bacterial cultures.

Gas Formation. Many bacteria, anaërobic ones especially, produce gases, noxious and odorless; in the culture-media the bubbles which arise soon displace the media.

Odors. Some germs form odors characteristic of them: some sweet, aromatic ones, and others very foul, disagreeable smells; some give a sour or rancid exhalation.

Effect of Age. With age, bacteria lose their strength and die.

Bacteria thus carry on all the functions of higher organized life; they breathe, eat, digest, excrete, and multiply; and they are very busy workers.

CHAPTER III.

METHODS OF EXAMINATION.

We divide the further study of the general characteristics of Bacteria into two portions :—

First the *examination of the same* by aid of the microscope.

Second. The continued study *through artificial cultivation.*

They both go hand in hand ; the one incomplete without the other.

Microscopical. The ordinary microscope will not suffice for Bacteriological research. Certain special appliances must first be added. It is not so much required to have a picture very large, as to have it sharp and clear.

Oil Immersion Lens. The penetration and clearness of a lens are very much influenced by the absorption of the rays of light emerging from the picture. In the ordinary dry system, many of the light rays, being bent outward by the air which is between the object and the lens, do not enter the lens, and are lost. By interposing an agent which has the same refractive index as glass, *cedar-oil, or clove-oil,* for example, all the rays of light from the object enter directly into the lens.

The "Homogeneous System," as this lens is called, dips into a drop of cedar-oil placed upon the cover-glass, and is then ready for use.

Abbe's Condenser. The second necessary adjunct is a combination of lenses placed underneath the stage, for bringing wide rays of light directly under the object. It serves to intensify the colored pictures by absorbing or hiding the unstained structure.

Fig. 7.

Abbe's Condenser.

This is very useful in searching a specimen for bacteria, since it clears the field of everything that is not stained. It is called Abbe's Condenser. Together with it is usually found an instrument for shutting off part of the light—a *Blender*. When the bacteria have been found, and their relation to the structure is then wished to be studied, the "Abbe" is generally shut out by the Iris blender, and the structure comes more plainly into view.

Fig. 8.

Iris Blender.

For all *stained Bacteria* the oil immersion lens and Abbe condenser, without the use of Blender. For *unstained specimens*, oil immersion and the narrowed blender.

When examining with low power objective, use a *strong* ocular. When using high power objective use *weak* ocular. A nose-piece will be found very useful, since it is sometimes necessary to change the objective on the same field, and that insures a great steadiness of the object.

Great cleanliness is needed in all bacteriological methods ; but nowhere more so than in the microscopical examination.

The cover-glass should be very carefully washed in alcohol, and dried with a soft linen rag. To remove the stains on the cover-glasses that have been used, they should be soaked in hydrochloric acid.

They are well cleaned by cooking fifteen minutes in a ten per cent. lysol solution.

Examination of Unstained Bacteria. As the coloring of bacteria kills them and changes their shape to some extent, it is preferable to examine them when possible in their natural state.

We obtain the bacteria for examination, either from liquid or solid media.

From Liquids. With a long platinum needle, the end of which is bent into a loop, we obtain a small drop from the liquid containing the bacteria, and place it on a cover-glass or slide; careful that no bubbles remain.

Sterilize Instruments. Right here we might say that it is best to accustom one's self to passing all instruments, needles, etc., through the flame before and after each procedure; it insures safety; and once in the habit, it will be done automatically.

From Solid Media. With a straight-pointed platinum needle, a small pinch of the medium is taken and rubbed upon a glass slide, with a drop of sterilized water, or bouillon, and from this a little taken on cover-glass, as before.

FIG. 9.

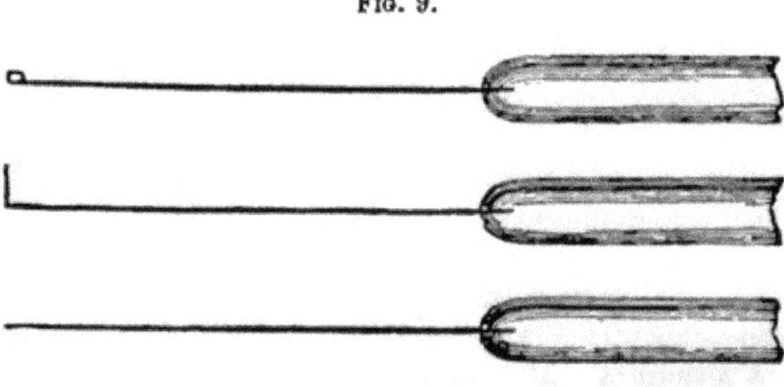

Platinum Needles.

The cover-glass with its drop is now placed on the glass slide, carefully pressing out all bubbles. Then a drop of cedar-oil is laid on top of the cover-glass, and the oil immersion lens dipped gently down into it as close as possible to the cover-glass, the narrow blender *shutting off* the Abbe condenser, for this being an unstained specimen, we want but *little light.* We now apply the eye, and if not in focus, use the fine adjustment, or, using the coarse, but always *away from the object* that is towards us, since the distance between the specimen and the lens is very slight, it does not require much turning to break the cover-glass and ruin the specimen. Having found the bacterium, we see whether it be bacillus, micrococcus, or spirillum; discover if it be motile, or not. That is about all we can ascertain by this method.

Fig. 10.

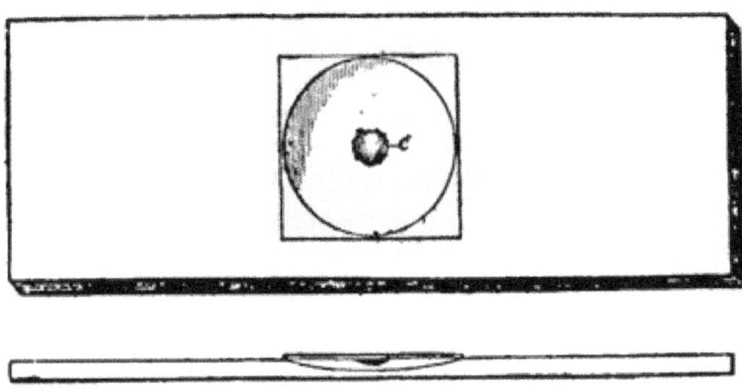

Hanging Drop in Concave Glass Slide.

Hanging Drop. When the looped platinum needle is dipped into a liquid, a very finely-formed globule will hang to it; this can be brought into a little cupped glass slide (an ordinary microscopic glass slide with a circular depression in the centre) in the following manner: The drop is first brought upon a cover-glass; the edges of the concavity on the glass slide are smeared with vaseline, and the slide inverted over the drop; the cover-glass sticks to the smeared slide, which, when turned over, holds the drop in the depression covered by the cover-glass, thus forming an air-tight cell; here the drop cannot evaporate.

Search for the bacteria with a weak lens; having found them, place a drop of cedar-oil upon the cover-glass, and bring the oil immersion into place (here is where a nose-piece comes in very usefully), careful not to press against the cell, for the cover-glasses are very fragile in this position.

Search the *edges* of the drop rather than the middle; it will usually be very thick in the centre and the bacteria not so easily distinguished.

Spores, automatic movements, fission, and cultivation in general can be studied for several days. This *moist chamber* can be placed in a brood-oven or on the ordinary warming stages of the microscope.

CHAPTER IV.

STAINING OF BACTERIA.

STAINING or coloring bacteria is done in order to make them prominent, and to obtain permanent specimens. It is also necessary to bring out the structure of the bacteria, and serves in many instances as a means of diagnosis; and lastly, it would be well-nigh impossible to discover them in the tissues, without staining.

Only since the *aniline* colors have come into active use, has the technology of staining become developed.

Aniline Colors. Of the numerous dyes in the market, nearly all have, at one time or other, been used in staining bacteria. But now only a very few find general use, and with methyline blue and fuchsin nearly every object can be accomplished.

Basic and Acid Dyes. Ehrlich was the first to divide the aniline dyes into two groups, the basic colors to which belong—

Gentian violet,
Methyl violet,
Methylin blue,
Fuchsin,
Bismark-brown,

And the acid colors to which *eosin* belongs.

The *basic* dyes stain the bacteria and the nuclei of cells; the *acid* dyes stain chiefly the tissue, leaving the bacteria almost untouched. *Carmine* and *Hæmatoxylin* are also useful as contrast stains, affecting bacteria very slightly. The aniline dyes are soluble in alcohol or water or a mixture of the two.

Staining Solutions. A saturated solution of the dye is made with alcohol. This is called the *stock* or *concentrated* solution;

1 part of this solution to about 100 parts of distilled water constitutes the ordinary aqueous solution in use or *weak* solution.

It is readily made by adding to an ounce bottle of distilled water enough of the strong solution until the fluid is still opaque in the body of the bottle, but clear in the neck of the same.

These weak solutions should be *renewed* every three or four weeks, otherwise the precipitates formed will interfere with the staining.

Compound Solutions. By means of certain chemical agents, the intensity of the aniline dyes can be greatly increased.

Mordants. Agents that "*bite*" into the specimen carrying the stain with them, depositing it in the deeper layers, are called mordants or etchers.

Various metallic salts and vegetable acids are used for such purpose.

The mother liquid of the aniline dyes, *aniline oil*, a member of the aromatic benzol group, has also this property.

Aniline Oil Water. Aniline oil is shaken up with water and then filtered; the aniline water so obtained is mixed with the dyes forming the "aniline water gentian violet" or aniline water fuchsin, etc.

Carbol Fuchsin. Carbolic acid can be used instead of aniline oil, and forms one of the main ingredients of Ziehl's or Neelsen's solution, used principally in staining bacillus tuberculosis. Kühne has a carbol-methylin blue made similar to the carbol fuchsin.

Alkaline Stains. Alkalies have the same object as the above agents; namely, to intensify the picture. Potassium hydrate, ammon. carbonate, and sodium hydrate are used.

Löffler's alkaline blue and Koch's weak alkaline blue make use of potassium.

Heat. Warming or boiling the stains during the process of staining increases their intensity.

Decolorizing Agents. The object is usually over-colored in some part, and then *decolorizing* agents are employed. Water is sufficient for many cases; alcohol and strong mineral acids combined are necessary in some.

Iodine as used in Gram's Method. Belonging to this group, but used more in the sense of a protective, is *tr. iodine.* It picks out certain bacteria, which it coats; prevents *them* from being decolorized, but allows all else to be faded. Then by using one of the acid or tissue dyes, a contrast color, or double staining is obtained. Many of the more important bacteria are not acted upon by the iodine, and it thus becomes a very useful means of diagnosis.

Formulas of different Staining Solutions.

I.—*Saturated Alcoholic Solution.*

Place about 10 grammes of the powdered dye in a bottle and add 40 grammes of alcohol. Shake well and allow to settle. This can be used as the stock bottle.

II.—*Weak Solutions.*

Made best by adding about 1 part of number I. or stock solution to 10 of distilled water. This is the ordinary solution in use.

III.—*Aniline Oil Water.*

Aniline oil	5 parts.
Distilled water	100 parts.—M.

Shake well and filter. To be made fresh each time.

IV.—*Aniline Water Dyes.*

Sat. alcoh. sol. of the dye . . .	11 parts.
Aniline oil water	100 parts.
Abs. alcohol	10 parts.—M.

Can be kept 10 days.

V.—*Alkaline Methylin Blue.*

A. *Löffler's.*

Sat. alc. sol. methylin blue . .	30
Sol. potass. hydrat. (1-10,000) .	100—M.

B. *Koch's.*

Sol. potass. hydrat. (10 per cent.)	0.2
Sat. alc. sol. methyl. blue . . .	1.0
Distilled water	200.0—M.

VI.—*Carbolic Acid Solutions.*

A. *Ziehl-Neelsen.*

Fuchsin (powd.)	. . . .	1 part.
Alcohol	. . . .	10 parts.
5 per cent. sol. acid. carbolic	.	100 parts.—M.

Filter. The older the solution the better.

B. *Kühne.*

Methylin blue	. . . . .	1.5
Alcohol	. . . .	10.0
5 per cent. sol. ac. carbol.	. .	100.0

Add the acid gradually. This solution loses strength with age.

VII.—*Gram's Iodine Solution.*

Iodine	. . . .	1
Potass. iod.	. . . . . .	2
Aquæ destillat.	. . . .	300.—M.

VIII.—*Löffler's Mordant* (for flagella).

Aq. sol. of tannin (20 per cent.)	.	10 parts.
Aq. sol. ferri sulph. (5 per cent.)	.	1 part.
Aquæ decoc. of logwood (1-8)	.	4 parts.—M.

Keep in well-corked bottle.

IX.—*Picro-carmine* (Ranvier).

Carmine	. . . . . .	1
Water	. . . .	10
Sol. ammon.	. . . . . .	3
Sat. sol. picric acid	. . . .	200.—M.

X.—*Gabbet's Acid Blue* (rapid stain).

Methylin blue	. . . . .	2
25 per cent. sulphuric acid	. .	100.—M.

XI.—*Alkaline Aniline Water Solutions.*

Sodium hydrat. (1 per cent.)	. .	1
Aniline oil water	. . . . .	100.—M.

And add—

Fuchsin, or methyl-violet powd.	.	4

Cork well. Filter before using.

CHAPTER V.

GENERAL METHOD OF STAINING SPECIMENS.

Cover-Glass Preparations. The material is evenly spread in as thin a layer as possible upon a cover-glass; then, to spread it still more finely, a second cover-glass is pressed down upon the first and the two slid apart. This also secures two specimens. Before they can be stained they must be perfectly dry, otherwise deformities will arise in the structure.

Drying the Specimen.—The cover-glass can be set aside to dry, or held in the fingers over the Bunsen burner (the fingers preventing too great a degree of heat). Since most of the specimens contain a certain amount of albumenoid material, it is best in all cases to "fix" it, *i. e.*, to coagulate the albumen. This is accomplished by passing the cover-glass (after the specimen is dry) three times through the flame of the burner, about three seconds being consumed in doing so, the glass being held in a small forceps, smeared side up.

The best forceps for grasping cover-glasses is a bent one, bent again upward, near the ends. (Fig. 11.) It prevents the flame or staining-fluid from reaching the fingers.

FIG. 11.

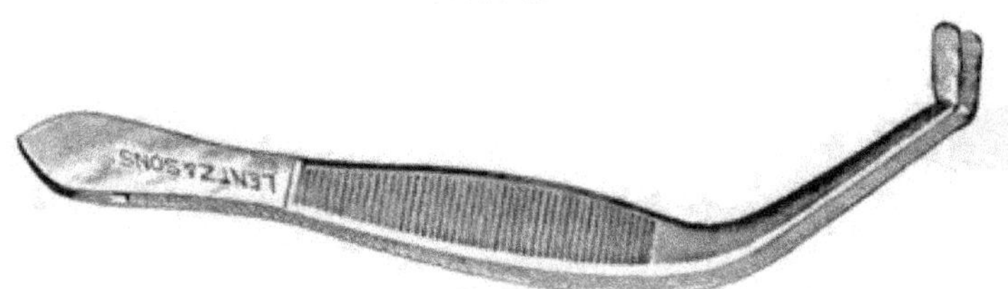

Author's Bent Forceps for Holding Cover-glass over Flame.

The object is now ready for staining.

Staining.—A few drops of the staining solution are placed upon the cover-glass so that the whole specimen is covered, and it is left on a few minutes, the time depending upon the variety, the strength of stain, and the object desired. Instead of placing the dye upon the object, the cover-glass can be immersed in a small glass dish containing the solution; or, if heat is desired to intensify or hasten the process, a watch-crystal holding the stain is placed over a Bunsen burner and in it the cover-glass; and, again, the cover-glass can be held directly in the flame with the staining fluid upon it, which must be constantly renewed until the process is completed.

Removing Excess of Stain. The surplus stain is washed off by dipping the glass in water, distilled water always best, though ordinary running water is admissible.

The water is *removed* by drying between filter paper or simply allowed to run off by standing the cover-glass slantwise against an object. When the specimen is to be examined in water (which is always best with the first preparation of the specimen, as the Canada balsam destroys to some extent the natural appearance of the bacteria), a small drop of sterilized water is placed upon the glass slide, and the cover-glass dropped gently down upon it, so that the cover-glass remains adherent to the slide.

The dry system or the oil-immersion can now be used.

When the object has been sufficiently examined it can be *permanently* mounted by lifting the cover-glass off the slide (this is facilitated by letting a little water flow under it, one end being slightly elevated). The water that still adheres is dried off in the air or gently over the flame, and when perfectly dry it is placed upon the drop of Canada balsam which has been put upon the glass slide.

In placing the cover-glass in the staining solutions one must be careful to remember which is the spread side.

By holding it between yourself and the window, and scraping the sides carefully with the sharp point of the forceps, the side having the specimen on it will show the marks of the instrument.

Little glass dishes, about one-half-dozen, should be at hand for containing the various stains and decolorants.

Tissue Preparations. In order to obtain suitable specimens for staining, very thin sections of the tissue must be made.

As with histological preparations, the tissue must be hardened before it can be cut thin enough. Alcohol is the best agent for this purpose.

Pieces of the tissue one-quarter inch in size are covered with alcohol for 24 to 48 hours.

When hardened it must be fixed upon or in some firm object. A paste composed of—

Gelatine	1 part.
Glycerine	4 parts.
Water	2 parts.

will make it adhere firmly to a cork in about 2 hours, or it can be imbedded in a small block of paraffine, and covered over with melted paraffine.

Cutting. The microtome should be able to cut sections $\frac{1}{5000}$ inch in thickness; this is the fineness usually required.

The sections are brought into alcohol as soon as cut unless they have been imbedded in paraffine, when they are first washed in chloroform to dissolve out the paraffine.

Staining. All the various solutions should be in readiness, best placed in the little dishes in the order in which they are to be used, as a short delay in one of the steps may spoil the specimen.

Fig. 12.

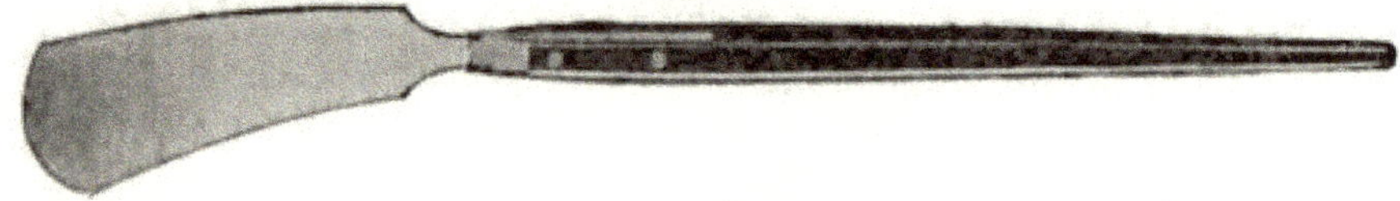

Spatula for Lifting Sections.

A very useful instrument for transferring the delicate sections from one solution to another is a little metal spatula, the blade being flexible.

A still better plan, especially when the tissue is "crumbling," is to "carry out" the whole procedure on the glass-side.

General Principles. The section is transferred from the alcohol in which it has been kept into water, which removes the excess of alcohol, from here into—

Dish I, containing the *stain;* where it remains 5 to 15 minutes. Then—

Dish II, containing *5 per cent. acetic acid* (1 to 20); where it remains ½ to 1 min. The acid removes the excess of stain.

Dish III, *water* to rinse off the acid. The section can now be placed under the microscope covered with cover-glass to see if the intensity of the stain is sufficient or too great. A second section is then taken, avoiding the errors, if any; and having reached this stage proceeded with as follows:—

Dish IV, *alcohol*, 2 to 3 seconds to remove the water in the tissue.

V. A few drops of *oil of cloves*, just long enough to clear the specimen to make it transparent (so that an object placed underneath will shine through).

VI. Remove excess with filter-paper.

VII. Mount in Canada balsam.

FIG. 13.

Section Microtome.

CHAPTER VI.

SPECIAL METHODS OF STAINING AND MODIFICATIONS.

Gram's Method of Double Staining. (For cover-glass specimens.)—I. A hot solution of anil. water gentian violet 2 to 10 minutes.

II. Directly without washing, into Gram's solution of iod. potass. iod. 1 to 3 min. (the cover-glass looks black).

III. Wash in alcohol 60 per cent. until only a light brown shade remains (as if the glass were smeared with dried blood).

IV. Rinse off alcohol with water.

V. Contrast color with either eosin, picro-carmine, or bismarck-brown. The bacteria will appear deep blue, all else red or brown on a very faint brown background.

The following bacteria do not retain their color with Gram's method—are therefore not available for the stain :—

Bacillus of typhoid.
Spirillum of cholera.
Bacillus of chicken cholera.
" of hemorrhagic septicæmia.
" of malignant œdema.
" of pneumonia (Friedlander).
" of glanders.
Diplococcus of gonorrhœa.
Spirillum of relapsing fever.

Gram's Method for Tissues (modified by Günther).

I. Stain in anil. water gent. violet . . 1 minute.
II. Dry between filter paper.
III. Iod. potass. iod. sol. . . . . . 2 minutes.
IV. Alcohol . . . . ½ minute.
V. 3 per ct. sol. hydrochloric acid in alcohol 10 seconds.
VI. Alcohol, ol. of cloves, and Canada balsam.

To Stain Spores. Since spores have a very firm capsule, which tends to keep out all external agents, a very intensive stain is required to penetrate them, but once this object attained it is equally as difficult to decolorize them.

A cover-glass prepared in the usual way, *i. e.*, drying and passing the specimen through the flame three times, is placed in a watch-crystal containing Ziehl's carbol-fuchsin solution, and the same placed upon a rack over a Bunsen burner, where it is kept at boiling-point for *one hour*, careful to supply fresh solution at short intervals lest it dry up.

The bacilli are now decolorized in alcohol, containing ½ per cent. hydrochloric acid. A contrast color, preferably methylin blue, is added for a few minutes.

The spores will appear as little red beads in the blue bacteria, and loose ones lying about.

Spore Stain (modified).—I. *Carbol.-fuchsin* on cover-glass and heated in the flame to boiling point 20 to 30 times.

II. 25 per cent. sulphuric acid, 2 seconds ; rinsed in water.

III. Methylin blue contrast.

Flagella Stain, *with Löffler's Mordant.*—I. A few drops of the mordant (No. viii.) are placed upon the spread cover-glass and heated until it steams.

II. Washed with water until the cover-glass looks almost clean, using a small piece of filter paper to rub off the crusts which have gathered around the edges.

III. Aniline water fuchsin (neutral) held in flame about 1½ minutes.

IV. Wash in water.

If the stain is properly made, the microbes are deeply colored and the flagella seen as little dark lines attached to them.

Sporogenic bodies stain quite readily, and in order to distinguish them from spores *Ernst* uses *alkaline methylin* blue, slightly warmed.

Then rinse in water.

Contrast with cold bismark-brown.

The spores are colored bright blue, the spore granules a dirty blue, being mixed with the brown, which colors also the bacteria.

Kühne's Method.—In sections, the alcohol used sometimes decolorizes too much. To obviate this *Kühne* mixes the alcohol with the stain, so that while the section is being anhydrated it is constantly supplied with fresh dye.

Weigert uses aniline oil to dehydrate instead of alcohol, and here, too, it can be used mixed with the dye.

General Double Staining for Sections.

I.	Stain (watery dyes)	10 to 15 minutes.
II.	Acetic acid and water (1 to 4) .	½ minute.
III.	Alcohol	2 to 3 minutes.
IV.	Contrast stain, usually picro-carmine or eosin	2 to 3 minutes.
V.	Alcohol	½ minute.
VI.	Clove oil. Canada balsam.	

Instead of coloring with the contrast last, it can be used first, then alcohol one-half minute, followed by the bacteria stain, acid water, alcohol, clove oil, and Canada balsam in succession.

The stains for special bacteria will be given when treating of the same.

CHAPTER VII.

METHODS OF CULTURE.

Artificial Cultivation.—The objects of cultivation are to obtain germs in pure culture, free from all foreign matter, isolated and so developed as to be readily used either for microscopical examination or animal experimentation.

To properly develop bacteria we supply as near as possible the conditions which hold for the especial germ in nature. With the aid of solid nutrient media the bacteria can be easily separated, and the methods are nearly perfect.

Sterilization. If we place our nutrient material in vessels that have not been properly disinfected, we will obtain growths of bacteria without having sown any.

If we have thoroughly cleaned our utensils, and then not taken care to protect them from further exposure, the germs we have sown will be effaced or contaminated by multitudes of others, that are constantly about us. We therefore have two necessary precautions to take :—

First. To thoroughly clean and sterilize every object that enters into, or in any way comes in contact with, the culture,

Second. To maintain this degree of disinfection throughout the whole course of the growth, and prevent, by proper containers, the entrance of foreign germs.

Disinfectants. Corrosive sublimate (bichloride of mercury), which is the most effective agent we possess, cannot be generally used because it renders the soil unproductive and therefore must only be employed in washing dishes, to destroy the old cultures. Even after washing, a few drops of the solution may remain and prevent growth, so that one must be careful to have the glass-ware that comes in contact with the nutrient media not too moist with the sublimate.

Heat. Heat is the best agent we possess for general use. Dry heat and moist heat are the two forms employed.

Fig. 14.

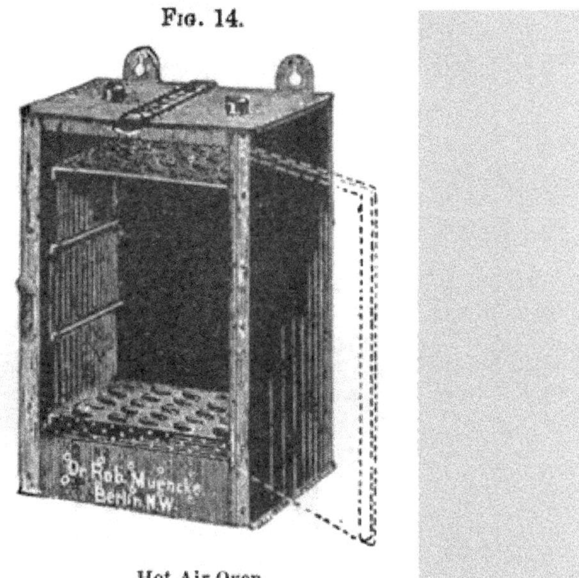

Hot Air Oven.

For obtaining *dry heat*—that is, a temperature of 150° C., (about 300° F.)—a sheet-iron oven is used which can be heated by a gas-burner. If it have double walls (air circulating between), the desired temperature is much more quickly obtained. A small opening in the top to admit a thermometer is necessary. These chests are usually about 1 foot high, 1¼ foot wide, and ¾ foot deep. In them, glassware, cotton, and paper can be sterilized. When the cotton is turned slightly brown, it usually denotes sufficient sterilization. All instruments, where practicable, should be drawn through the flame of an alcohol lamp or Bunsen burner.

Moist Heat.—Steam at 100° C, in circulation has been shown to be a very effective application of heat.

Koch's Steam-chest. Circulating steam is obtained by aid of Koch's apparatus. This consists of a cylindrical tin chest

FIG. 15.

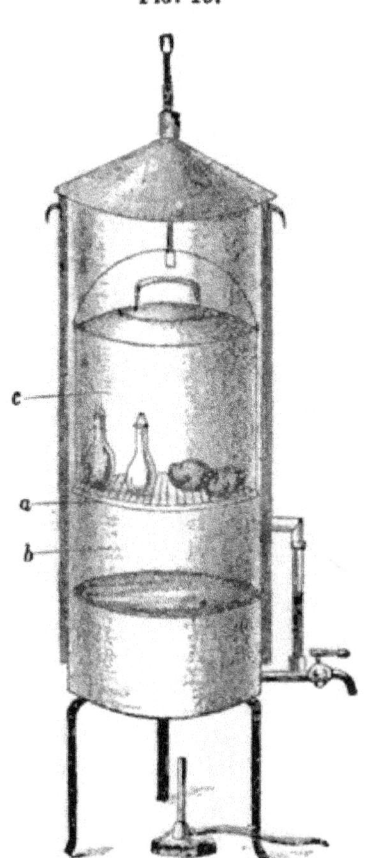

Koch's Steam-chest.

about 2½ feet high and about ½ foot in diameter; divided in its interior by a perforated diaphragm, *a*, an upper chamber for the steam, *c*, and a lower one for water, *b*. Two or more gas-burners placed underneath the chest, which stands on a tripod, supply the heat. In the cover is an opening for a thermometer. The chest is usually covered with felt. When the thermometer registers 100° C. the culture-medium or other substance to be sterilized is placed in the steam and kept there from 10 to 15 minutes, or longer, as required.

Arnold's Steam-sterilizer will answer every purpose of the Koch steam-chest. It is cheaper, requiring also less fuel to keep it going. The steam does not escape, but is condensed in the outer chamber. (Fig. 16.)

Fig. 16.

Arnold's Steam-sterilizer.

The autoclave of Chamberland allows a temperature of 120° C. to be obtained, and is much used in Pasteur's laboratory.

Instead of sterilizing for a long time at once, successive sterilization is practised with nutrient media, so that the albumen will not be too strongly coagulated. Fifteen minutes each day for three days in succession.

Fig. 17.

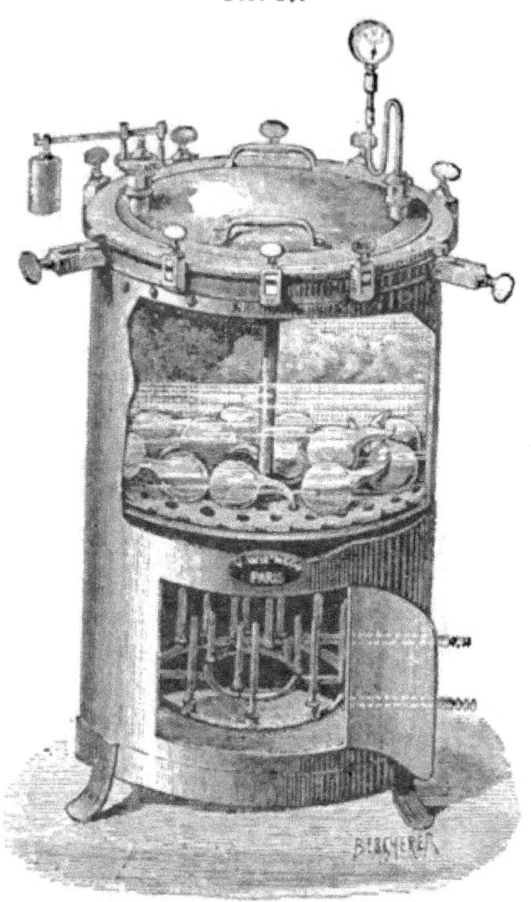

Chamberland's Autoclave with pressure.

Fractional Sterilization of Tyndall. Granted that so many spores originally exist in the object to be sterilized, it is subjected to 60° C. for four hours, in which time a part at least of those spores have developed into bacteria, and the bacteria destroyed by the further application of the heat. The next day more bacteria will have formed, and four hours' subjection to 60° heat will destroy them, and so at the end of a week, using four hours' application each day, all the spores originally present will have germinated and the bacteria destroyed.

Cotton Plugs or Corks. All the glass vessels (test-tubes, flasks, etc.) must be closed with cotton plugs, the cotton being easily sterilized and preventing the entrance of germs.

Test-tubes. New test-tubes are washed with hydrochloric acid and water to neutralize the alkalinity often present in fresh glass. They are then well washed and rubbed with a brush, placed obliquely to drain, and when dry corked with

Fig. 18. Fig. 19.

Wire-Cage.

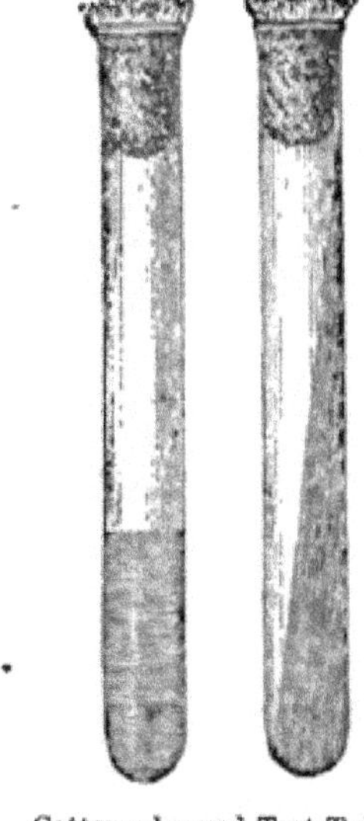

Cotton plugged Test-Tubes.

cotton plugs. Then put in the hot-air oven (little wire-cages being used to contain them) for fifteen minutes, after which they are ready to be filled with the nutrient media. (The cotton should fit firmly in the tube and extend a short space beyond it.)

Test-tubes without flaring edges are more desirable since the edges can easily be drawn out so as to seal the tube.

Instead of test-tubes, ordinary 3 oz. panel medicine bottles can be used for retaining the nutrient media and cultures.

According to late investigations, the glass tubes become sufficiently sterile in the steam-chest without the preliminary sterilization in the dry oven.

CHAPTER VIII.

NUTRIENT MEDIA.

Of the many different media recommended and used since bacteriology became a science, we can only describe the more important ones now in use. Each investigator changes them according to his taste.

Fluid Media.

Bouillon (according to Löffler). A cooked infusion of chopped beef made slightly alkaline with carbonate of soda. Prepared as follows: 500 grammes of finely-chopped raw lean beef are placed in a wide-mouthed jar and covered with 1 litre of water; this is left standing twelve hours with occasional shaking. It is then strained through cheese cloth or straining cloth, the white meat remaining in the cloth being pressed until one litre of the blood red meat-water has been obtained. The meat-water must now be cooked, but before doing this, in order to prevent all the albumen from coagulating, 10 parts of peptone powder and 5 parts of common salt are added to every 1000 parts meat-water. It is next placed in the steam-chest or water-bath for three-quarters of an hour.

Neutralization. The majority of bacteria grow best on a neutral or slightly alkaline soil, and the bouillon, as well as other media, must be carefully neutralized with a sat. sol. of carbonate of soda. Since too much alkalinity is nearly as bad as none at all, the soda must be added drop by drop until red litmus paper commences to turn blue. The bouillon is then cooked another hour, and filtered when cold. The liquid thus obtained must be clearly alkaline, and not clouded by further cooking. If cloudiness occur, the white of an egg and further boiling will clear the same.

Sterilization of the bouillon. Erlenmeyer flasks (little conical glass bottles) or test-tubes plugged and properly sterilized are filled one-third full with the bouillon, and placed with their contents in the steam-chest. A tin pail with perforated bottom makes a good container in which they can be lowered in the Koch's oven. They are left in steam of 100° C. one hour for three successive days, after which the tubes and bouillon are ready for use.

Solid Media. The knowledge of bacteria and germs or moulds settling and growing upon slices of potato exposed to the air, led to the use of solid media for the artificial culture of the same. It was also thus learned that each germ tends to form a separate colony and remain isolated.

Potato-Cultures. A ripe potato with a smooth skin is the best.

Several are brushed and scrubbed with water to get rid of the dirt and the "eyes" are cut out.

Next placed in 1 to 500 solution of bichloride of mercury for ½ hour. Then in the steam-chest for ¾ hour.

In the meantime, a receptable is prepared for them. This is called the *moist chamber.*

The moist chamber consists of two large shallow dishes, one, the larger, as a cover to the other.

These dishes are washed in warm distilled water.

A layer of filter paper moistened with a 15 to 30 drops of 1 to 1000 bichloride is placed in the bottom of the glass dish.

Fig. 20.

Moist chamber for potatoes.

The operator now prepares his own hands, rolling up his coat sleeves and carefully washing his hands, then taking a potato from the steam-oven and holding it between his thumb and index finger in the short axis, he divides the potato in its long axis with a knife that has been passed through the flame. The two halves are kept in contact until they are lowered into the moist chamber, when they of their own weight fall aside, the cut surface uppermost. They are then ready for inoculation.

Fig. 21.

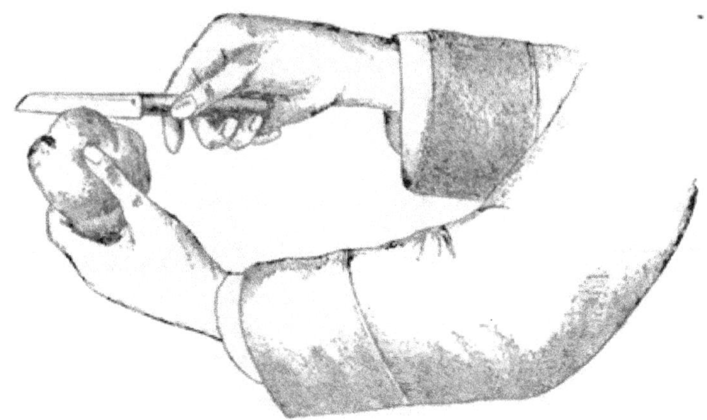

Method of slicing potato. (After Woodhead and Hare.)

Esmarch's Cubes. The potato is first well cleaned and peeled. It is then cut in cubes ½ inch in size.

These are placed, each in a little glass dish or tray and then in steam-chest for ½ hour, after which they are ready for inoculation (the dishes first having been sterilized in hot-air oven).

Test-tube Potatoes. Cones are cut out of the peeled potato and placed in test-tubes, which can then be plugged and easily preserved.

Manner of Inoculation. With a platinum rod or a spatula (sterilized) the material is spread out upon one of the slices, keeping free of the edges. The growth on this first or original potato will be quite luxuriant, and the individual colonies often difficult to recognize, therefore dilutions are made. (Fig. 22.)

From the original or first slice, a small portion including some of the meat of the potato is spread out upon the surface of a second slice, which is first dilution. From this likewise a small bit is taken and spread on a third slice or second dilution, and here usually the colonies will be sparsely enough settled to study them in their individuality.

This is the principle carried on in all the cultivations. It is a physical analysis.

FIG. 22.

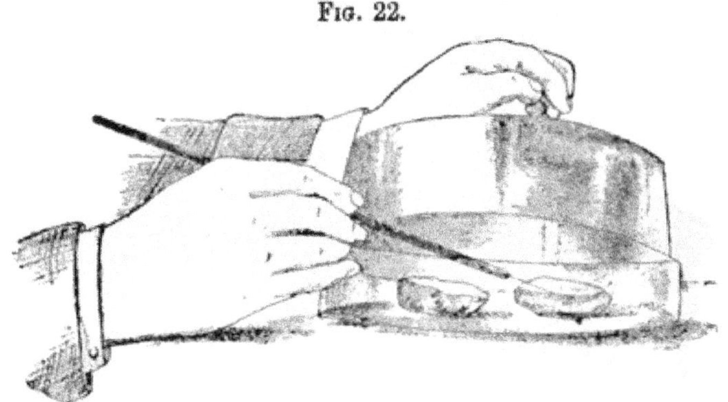

Method of inoculation. (Woodhead and Hare.)

Potato and Bread Mash. These pastes are used chiefly in the culture of moulds and yeasts. Peeled potatoes are mashed with distilled water until thick, and then sterilized in flasks ¾ of an hour for three successive days.

Bread Mush.—Bread devoid of crust, dried in an oven, and then pulverized and mixed with water until thick and sterilized as above.

CHAPTER IX.

SOLID TRANSPARENT MEDIA.

Solid Transparent Media are materials which can be used for microscopical purposes and which can readily be converted into liquids. Such are the gelatine and agar materials.

Gelatine. Gelatine is obtained from bones and tendons, and consists chiefly of chondrin and gluten.

The French golden medal brand is the one most in use, found in long leaves with ribbed lines crossing them.

Koch-Löffler 10 per cent. Bouillon-Gelatine. To the meat-water as made for the bouillon are added

100 grammes gelatine,
10 " peptone,
5 " salt,

to each 1000 grammes of the meat-water.

This is placed in a flask and gently heated until the gelatine is dissolved.

Neutralization with the soda and then cooking in water-bath for 1 hour or more until the liquid seems clear, then add white of an egg and boil ¼ hour longer; the egg will produce a clearer solution and save much trouble. A small portion, while hot, is now filtered into a test-tube and tested for alkalinity, and then re-heated several times, watching if a cloudy ppt. forms.

Fig. 23.

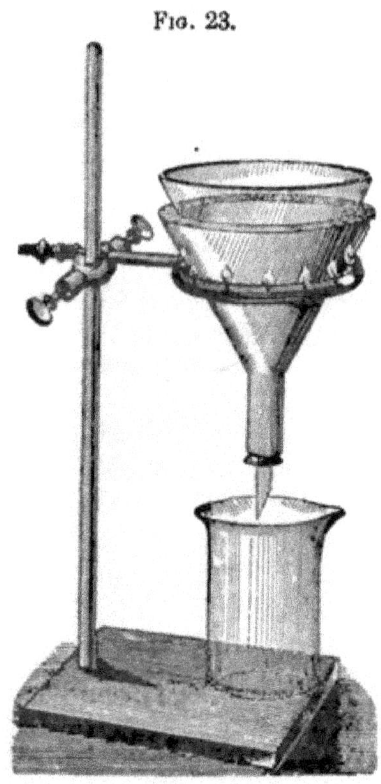

Hot-water filter.

If the fluid remains clear upon cooling, the remainder of the material can be filtered. It must be accomplished while hot, else the gelatine will coagulate and prevent further filtration.

This can be carried on either by keeping hot the solution continually in water-bath, and only filtering a small quantity at a time through the filter, or keeping the filter itself hot, either with a hot water filter or placing the filter in steam chest. (Fig. 23.)

Clouding of Gelatine. If the gelatine does not come out clear, or becomes turbid on cooling, it may be due to several things—

1. The filter-paper too thin or impure.
2. Too strongly alkaline.
3. Cooked too long or not long enough.

The addition of the white of an egg, as before mentioned, will often clear it up; if this avails not, re-filtering several times, and attention to the few points mentioned.

Sterilizing the Gelatine. The gelatine is kept in little flasks or poured at once into sterile test-tubes, careful not to wet the neck where the cotton enters, lest when cool the cotton plug stick to the tube.

The tubes are then placed in steam-chest for three successive days, 15 minutes each day (or in water-bath 1 hour a day for three days). Then set aside in a temperature of 15° to 20° C., and if no germs develop and the gelatine remains clear, it can be used for cultivation purposes.

Modifications. The amount of gelatine added to the meat-water can be variously altered, and instead of making gelatine bouillon the gelatine can be mixed with milk, blood, serum, urine, and agar-agar.

The nutrient gelatine bouillon can also receive additions in the shape of glycerine (4 per cent. to 6 per cent. being added), or reducing agents to take up the oxygen present.

Agar-Agar. This agent, which is of vegetable origin, derived from sea-plants gathered on the coasts of India and Japan, has many of the properties of gelatine, retaining its solidity at a much higher temperature; it becomes liquid at 90° C. and congeals again at 45° C. Gelatine will liquefy at 35° C.

It is not affected very much by the peptonizing action of the bacteria.

Preparation of Agar-Agar Bouillon or Nutrient Agar. The ordinary bouillon is first made, and then the agar cut in small pieces, added to the bouillon (15 grammes of agar to 1000 grammes bouillon).

It is allowed to stand several minutes until the agar swells, and then placed in water-bath or steam-chest for six hours or more. The reaction is taken, very little of the alkali being sufficient to neutralize it.

A white of an egg added, and boiled for several hours longer, when, even if not perfectly clear, it is filtered.

The filtering process, very difficult because of the readiness with which the agar solidifies, must be done in steam-chest or with hot-water filter, and very small quantities passed through at a time, changing the filter-paper often.

Cotton can be used here instead of filter-paper, or filtering entirely dispensed with by making use simply of decantation.

As agar is seldom clear, a little more or less opaqueness will not harm. The test-tubes are filled as with the gelatine, and sterilized in the same manner. While cooling, some of the tubes can be placed in a slanting position, so as to obtain a larger surface to work upon.

Water of condensation will usually separate and settle at the bottom, or a little white sediment remain encysted in the centre; this cannot easily be avoided, nor *does it form any serious obstacle.*

The crude agar should first be rinsed in clear water, and then in five per cent. acetic acid and clear water again, to rid it of impurities.

Glycerine Agar. The addition of 4 per cent. to 6 per cent. of glycerine to nutrient agar greatly enhances its value as a culture medium.

Gelatine-Agar. A mixture of 5 per cent. gelatine and 0.75 per cent. agar combines in it some of the virtues of both agents.

Blood Serum. Blood serum being rich in albumen coagulates very easily at 70° C., and if this temperature is not exceeded, a transparent, solid substance is obtained upon which the majority of bacteria develop, and some with preference.

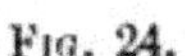
FIG. 24.

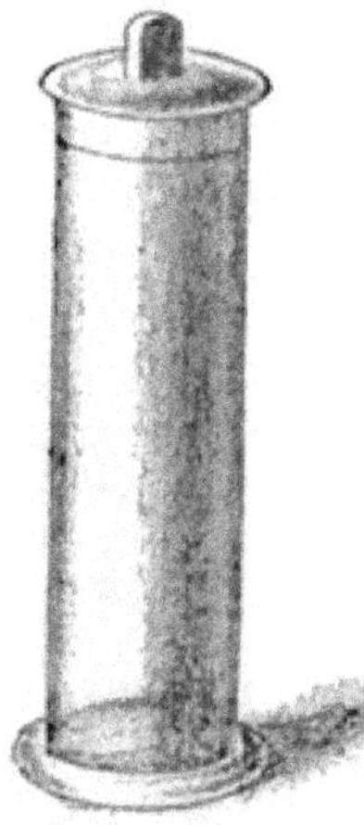
Flask to receive blood serum.

Preparation of Nutrient Blood Serum. If the slaughter of the animal can be supervised, it were best to have the site of the wound and the knife sterilized carefully, and then sterile flasks placed to receive the blood directly as it flows.

It is placed on ice forty-eight hours, and then the serum is drawn out with sterile pipettes into test-tubes; these are placed obliquely in an oven where the temperature can be controlled and maintained at a certain degree. See Fig. 25.

Incubators or Brood-ovens. Incubators or brood-ovens, as such ovens are called, consist essentially of a double-walled zinc or copper chest, the space between the walls filled with water.

The oven is covered with some impermeable material to prevent the action of surrounding atmosphere. (Fig. 26.) It is supplied with a thermometer and with a regulator. The regulator

Fig. 25.

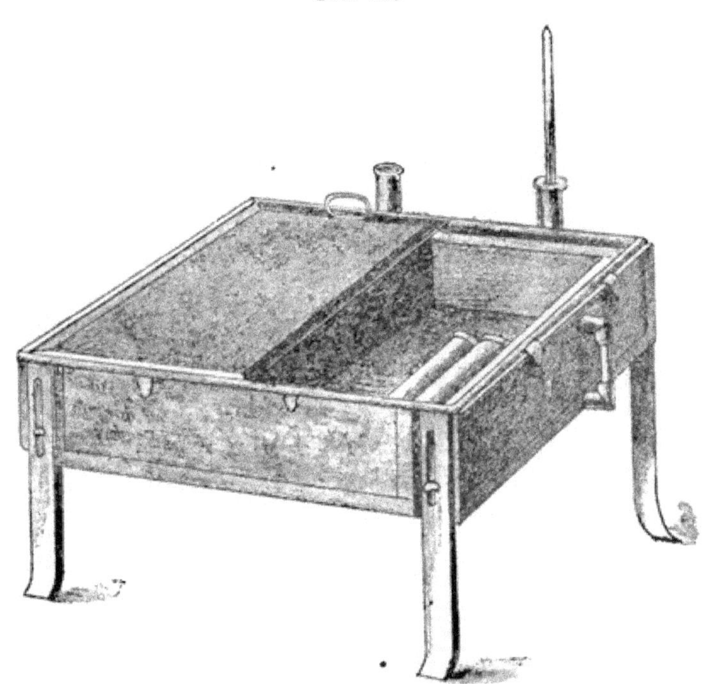

Thermostat for blood serum.

is connected with the Bunsen burner, and keeps the thermometer at a certain height.

There are several forms of regulators in use, and new ones invented continually.

The size of the flame in some is regulated by the expansion of mercury, which, as it rises, lessens the opening of the gas supply. The mercury contracting on cooling allows more gas to enter again. (Fig. 27.)

Koch has invented a *safety burner*, by which the gas supply is shut off should the flame accidentally have gone out.

Coagulation of Blood Serum. The tubes of blood serum having been placed in the oven, are kept at a temperature of 65° to 68° C., until coagulation occurs; then removed and sterilized.

FIG. 26. FIG. 27.

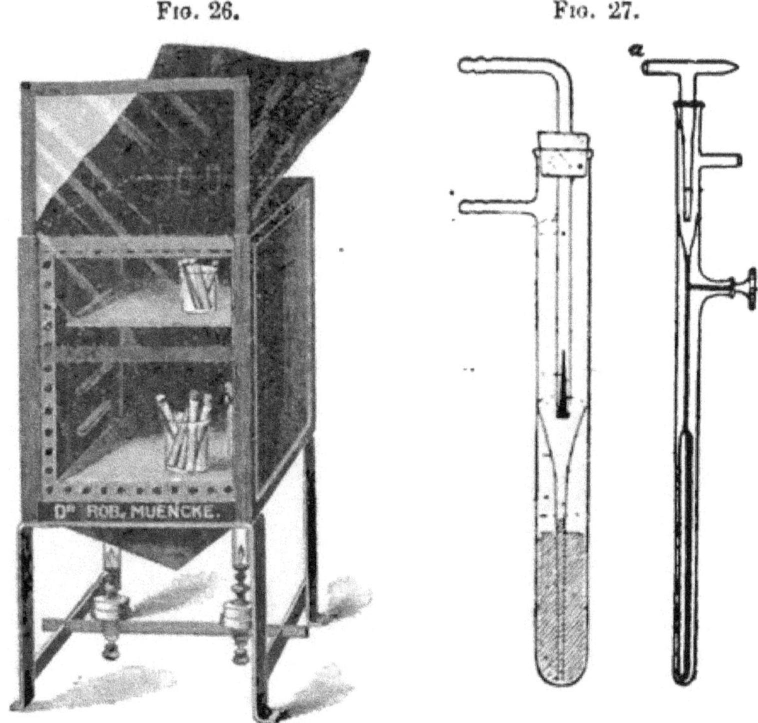

Babe's Incubator. Thermo-regulators.

Sterilization of Blood Serum. The tubes are placed 3 to 4 days in incubation at 58° C., and those tubes which show any evidences of organic growth are discarded.

If now, at the end of a week, the serum remains sterile at the ordinary temperature of the room, it can be used for experimental purposes.

Perfectly prepared blood serum is transparent, of a gelatine-like consistence, and straw-color. It will not liquefy by heat, though bacteria can digest it. Water of condensation always forms, which prevents the drying of the serum.

Blood serum, formerly much more used than now, was especially applicable to the culture of tubercle bacilli. The glycerine agar has now superseded it.

Human blood serum derived from placenta, serum from ascitic fluid, and ovarian cysts are prepared in a similar manner to the above.

Other Nutrient Media. Milk, urine, decoctions of various fruits and plants, and lately for cultivating anærobic bacteria, *eggs*.

Fresh Egg Cultures, after Hüppe. The eggs in the shell are carefully cleaned, washed with sublimate, and dried with cotton. The inoculation occurs through a very fine opening made in the shell with a hot platinum needle; after inoculation, the opening is covered with a piece of sterilized paper, and collodion over this.

Boiled Eggs. Eggs boiled, shell removed over small portion, and the coagulated albumen stroked with the material. (See *Diphtheria.*)

Guinea-pig Bouillon. The flesh of guinea-pigs as well as that of other experiment-animals is used instead of beef in the preparation of bouillon, for the growth of special germs.

CHAPTER X.

INOCULATION OF GELATINE AND AGAR.

Glass Slide Cultures. Formerly the gelatine was poured on little glass slides such as are used for microscopical purposes, and after it had become hard, inoculated in separate spots as with potatoes.

Test Tube Cultures. The gelatine, agar, or blood serum having solidified in an oblique position, is smeared on the surface with the material and the growth occurs, or the medium is punctured with a stab of the platinum rod containing the material. The first is called a *stroke* or *smear culture*, the second a *stab* or *thrust culture.* In removing the cotton plugs from the sterile tubes to carry out the inoculation, the plugs should remain between the fingers in such a way that the part which comes in contact with the mouth of the tube will not touch anything.

After the needle has been withdrawn the plugs are re-inserted and the tubes labelled with the kind and date of culture.

Plate Cultures. Several tubes of the culture medium are

made liquid by heating in water bath, and then inoculated with the material as follows. A looped platinum needle is dipped into the material and then shaken in the tube of liquid media, (gelatine, agar, etc.).

This first tube is called *original*. From this three drops (taken with the looped platinum rod) are placed in a second tube, the rod being shaken somewhat in the gelatine or agar; this is labeled *first dilution* (a colored pencil is useful for such markings).

Fig. 28.

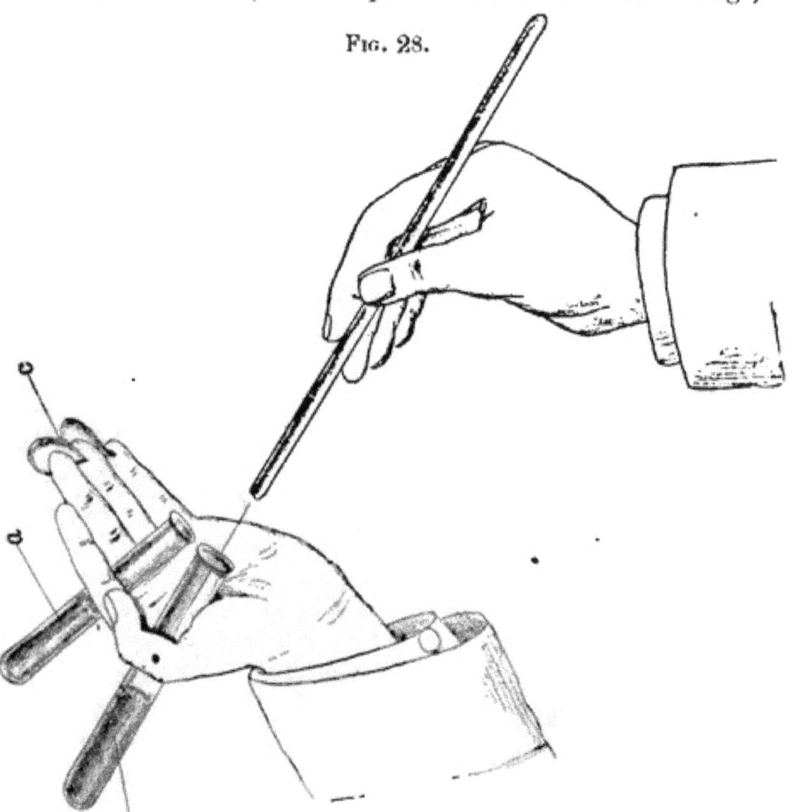

Manner of holding tubes for inoculation: *a*, tube with material; *b*, tube to be inoculated; *c*, cotton plugs. (After Woodhead and Hare.)

From the first dilution three drops are taken into a third tube, which becomes the *second dilution*. Fig. 28.

The plugs of cotton must be replaced after each inoculation, and during the same must be carefully protected from contamination.

To hasten the procedure and lessen the danger of contamination, the tubes can be held in one hand aside of each other, each

Fig. 29.

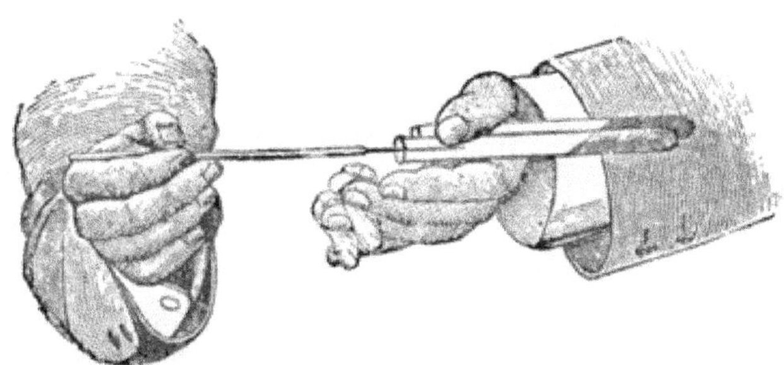

Manner of holding plugs.

plug opposite its tube. They are now ready for spreading on glass plates.

Glass Plates. The larger the surface over which the nutrient medium is spread the more isolated will the colonies be; window glass cut in rectangular plates 6x4 inches in size is used; about ten such plates are cleaned with dry towel and placed in a small iron box or wrapped in paper; and sterilized in the hot-air oven at a temperature of 150° C. for ten minutes. (Fig. 30.) When the plates have cooled they are placed upon an apparatus designed to cool and solidify the liquid media, which is now poured upon the plates from the inoculated test-tubes.

Fig. 30.

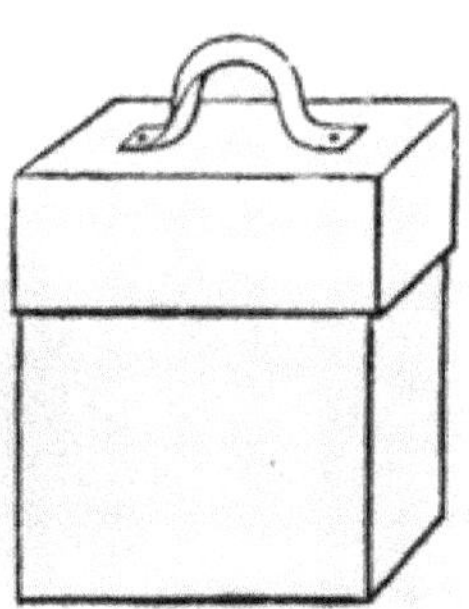

Iron box for glass plates.

Nivellier Leveling and Cooling Apparatus. Ice and water are placed in a shallow round glass tray; on top of this a square plate of glass, upon which the culture plate is placed, and covering this a bell-glass.

The whole is upon a low, wooden tripod, the feet of which can be raised or lowered, and a little spirit-level used to adjust it. (Fig. 31.) The glass plate taken out of the iron box is placed under the bell-glass. The tube containing the gelatine is held in the flame a second to singe the cotton plug to free it from dust, and the plug removed, the edges of the tube again flamed, the bell-glass lifted, and the inoculated gelatine carefully poured on the plate, leaving about one-third inch margin from the borders; the

Fig. 31.

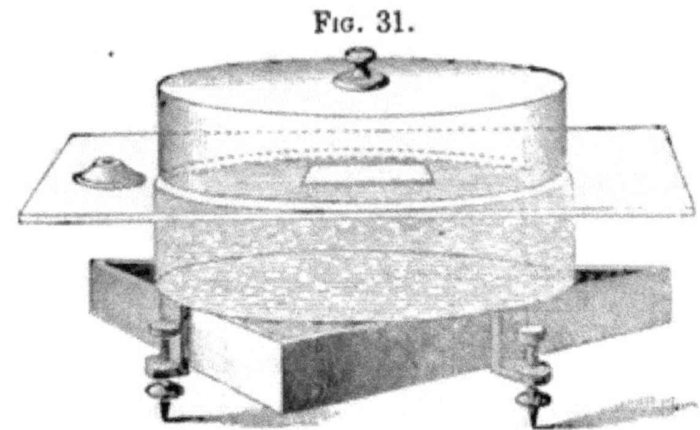

Nivellier leveling and cooling apparatus.

lips of the tube being sterile can be used to spread the media evenly. If the plate is at all cool, the fluid will solidify as it is being spread. The glass cover is replaced until the gelatine or agar is quite solid to prevent contamination.

Fig. 32.

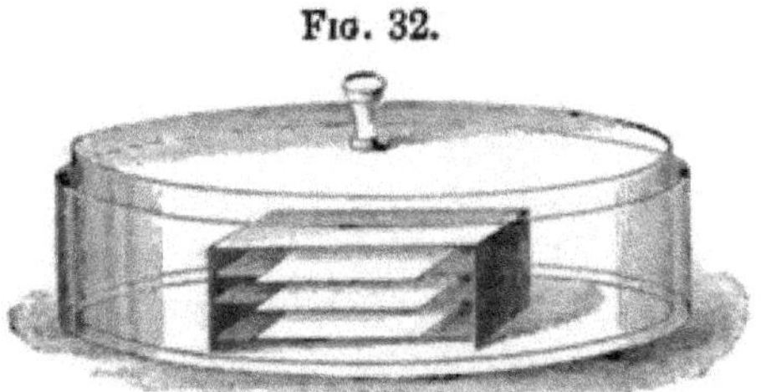

Moist chamber with plates on benches.

When the gelatine is congealed, the plate is placed upon a little glass bench or stand in the moist chamber.

The Moist Chamber Prepared Out of Two Glass Dishes, as for the Potato-Cultures. The glass benches are so arranged that one stands upon the other. In order to avoid confusion, a slip of paper with a number written on it is placed on the bench beneath each plate. As the original or first plate would have the colonies developed in greatest profusion, it is placed the first day on the topmost bench; but, since the colonies would be likely to overrun the plate and allow the gelatine to drop on the lower plates, it is best, as soon as evidences of growth appear, to place it below, and watch the third plate or second dilution for the characteristic colonies, forgetting not all this time to change the numbers accordingly.

The date of culture and the name can be written upon the moist chamber.

Petri Saucers. Agar hardens very quickly, even without any especial means for cooling, and it does not adhere very well to the glass. Therefore it is better to follow the method of Petri and use little shallow glass dishes, one covering the other. They are first sterilized by dry heat, and then the inoculated gelatine or agar is poured into the lower dish, covered by the larger one, and placed in some cool place, different saucers being used for each dilution.

FIG. 33.

Petri saucers.

This method is very useful for transportation; the saucers can be viewed under microscope similar to the glass plates, and has in a manner superseded them.

Esmarch's Tubes, or Rolled Cultures. This method, especially used in the culture of anærobic germs, consists in spreading the inoculated gelatine upon the inner walls of the test tube in which it is contained and allowing it to congeal. The colonies then develop upon the sides of the tube without the aid of other apparatus. The method is useful whenever a very quick and easy way is required. The rolling of the tube is done under ice-water or running water from the faucet. The tube is held a little slanting, so as to avoid getting too much gelatine around the cotton plug.

The tubes can be placed directly under the microscope for further examination of the colonies.

NOTE.—The peptone nutrient gelatine, blood serum and agar can now be purchased already prepared, thus saving a great deal of time and making unnecessary the purchase of considerable apparatus.

CHAPTER XI.

THE GROWTH AND APPEARANCES OF COLONIES.

Macroscopic. Depending greatly upon the temperature of the room, which should be about 65° C., the colonies develop so as to be visible to the naked eye in two to four days. Some require ten to fourteen days, and others grow rapidly, covering the third dilution in thirty-six hours. The plate should be looked at each day.

FIG. 34.

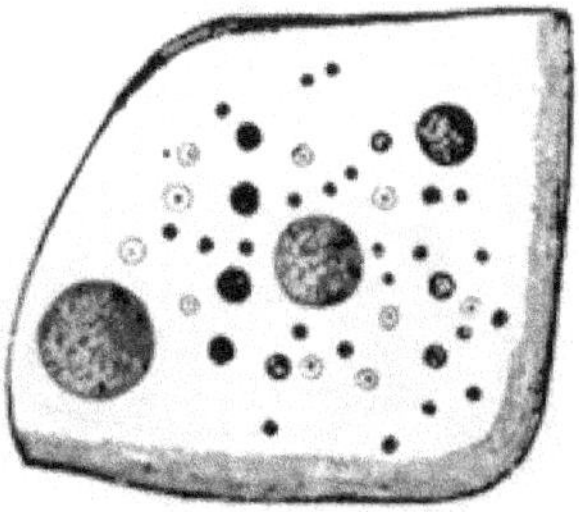

Naked eye appearances of colonies.

The colonies present various appearances, from that of a small dot, like a fly speck, to that resembling a small leaf. Some are elevated, some depressed, and some, like cholera, cup-shaped, umbilicated.

Then they are variously pigmented. Some liquefy the gelatine speedily, others not at all. The appearances of a few are so characteristic as to be recognized at a glance.

Microscopic. We use a low-power lens, with the abbé nearly shut out, that is the narrowest blender. The stage of the microscope should be of such size as to carry a culture plate easily upon it.

The second dilution or third plate is usually made use of, that one containing the colonies sufficiently isolated.

These isolated ones should be sought for, and their appearances well noticed.

There may be two or three forms from the same germ, the difference due to the greater or less amount of oxygen that they have received, or the greater or less amount of space that they have had to develop in.

The microscopic picture varies greatly; now it is like the gnarled roots of a tree, and now like bits of frosted glass; the pictures are very characteristic, and the majority of bacteria can be told thereby. (Fig. 31.)

Impression or "Klatsch" Preparations. In order to more thoroughly study a certain colony and to make a permanent specimen of the same, we press a clean cover-glass upon the particular colony, and it adheres to the glass. It can then be

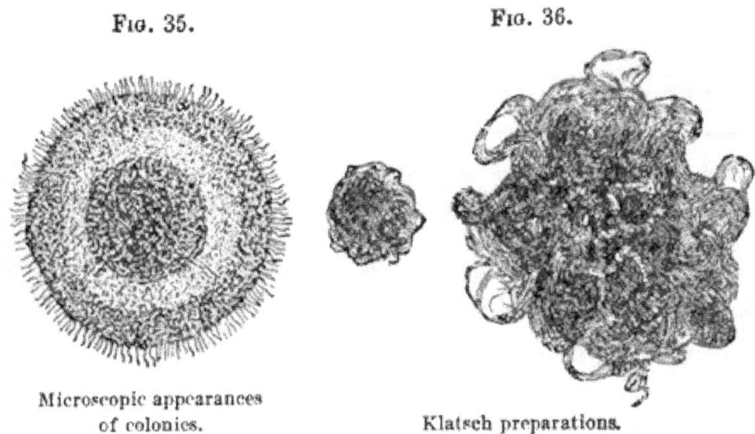

Fig. 35. Microscopic appearances of colonies.

Fig. 36. Klatsch preparations.

stained or examined so. The Germans give the name of "Klatsch" to such preparations. Many beautiful pictures can be so obtained.

Fishing. To obtain and examine the individual members of a particular colony the process of fishing, as it is called, is resorted to.

The colony having been placed under the field of the microscope, a long platinum needle, the point slightly bent, is passed between the lens and the plate so as to be visible through the microscope, then turned downward until the colony is seen to be disturbed, and the needle is dipped into the colony. This procedure must be carefully done, lest a different colony be disturbed than the one looked at, and an unknown or unwanted germ obtained.

After the needle has entered the particular colony, it is withdrawn, and the material thus obtained is further examined by staining and animal experimentation. The bacteria are then again cultivated by inoculating fresh gelatine, making *stab* and *stroke* cultures.

It is necessary to transfer the bacteria to fresh gelatine about every six weeks, lest the products of growth and decay given off by the organisms destroy them.

CHAPTER XII.

CULTIVATION OF ANÆROBIC BACTERIA.

SPECIAL methods are necessary for the culture of the anærobic variety of bacteria in order to procure a space devoid of oxygen. Several measures have been adopted of which the easiest and most serviceable are the following :—

FIG. 37.

Liborius's method.

Liborius's High Cultures. The tube is filled about ¾ full with gelatine, which is then steamed in a water bath and allowed to cool to 40° C., when it is inoculated by means of a long platinum rod with small loop, the movement being a rotary vertical one, and the rod going to the bottom of the tube.

The gelatine is next quickly solidified under ice; very little air is present. The anærobic germs will grow from the bottom upward, and any ærobins present will develop first on top, this method being one of isolation.

From the anærobic germs grown in the lower part, a stab culture is made into another tube containing ¾ gelatine, the material being obtained by breaking test-tube with the culture.

Hesse's Method. A stab culture having been made with anærobic germs, gelatine in a semi-solid condition is poured into the tube until it is full, thus displacing the air. (Fig. 38.)

Esmarch's Method. Having inoculated a tube with the microbe the gelatine is rolled out on the walls of the tube, a "roll culture," and the rest of the interior filled up with gelatine, the tube being held in ice water in the meanwhile. The colonies develop upon the sides of the tube and can be easily examined microscopically.

Gases like Hydrogen to replace the Oxygen. Several arrangements for passing a stream of hydrogen through the culture :—

Fränkel puts in the test tube, a rubber cork containing two glass tubes, one reaching to the bottom and connected with a hydrogen apparatus, the other very short, both bent at right angles. When the hydrogen has passed through ten to thirty minutes, the short tube is annealed and then the one in connection with the hydrogen bottle, and the gelatine rolled out upon the walls of the tube. (Fig. 39.) **Hüppe uses eggs as described in Chapter IX.**

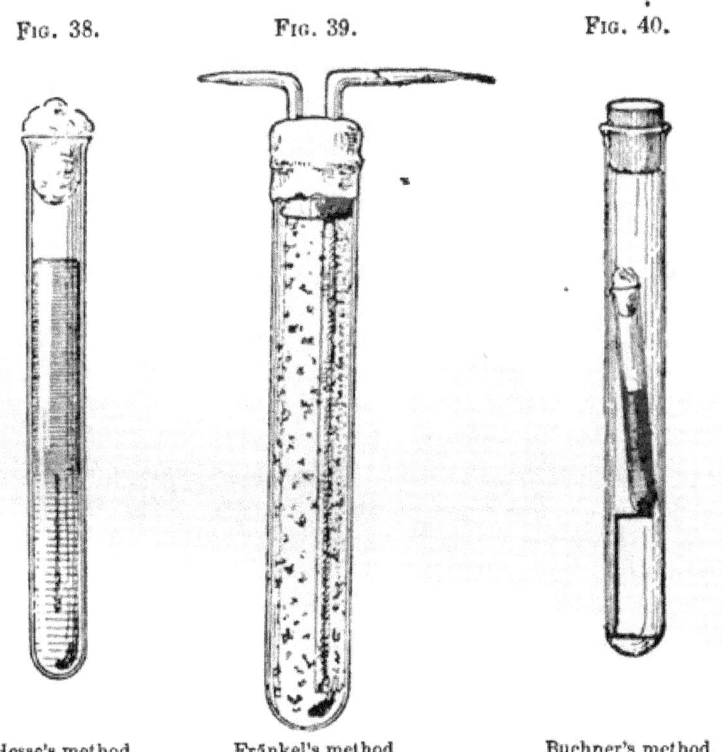

FIG. 38. Hesse's method.

FIG. 39. Fränkel's method.

FIG. 40. Buchner's method.

Use of Ærobic Bacteria to remove the Oxygen. Roux inoculates an agar tube through a needle thrust after which semisolid gelatine is poured in on top. When the gelatine has solidified, the surface is inoculated with a small quantity of bacillus subtilis or some other ærobic germ. The subtilis does not allow the oxygen to pass by, appropriating it to itself.

Buchner's Method. The test tube containing the culture is placed within a larger tube, the lower part of which contains an alkaline solution of pyrogallic acid or some other agent which absorbs oxygen. The tube is then closed with a rubber stopper. (Fig. 40.)

CHAPTER XIII.

THE WAY IN WHICH BACTERIA AFFECT THE ANIMAL ORGANISM.

BACTERIA affect the animal organism by depriving the cells of the body of oxygen and nitrogen which they appropriate to themselves for their maintenance.

They do more than this, however, for in their secretions and excretions the main potency lies.

Ptomaines, or *Cadaveric alkaloids*, was the name first applied to those bodies formed during putrefaction, but now used for all alkaloids or bodies of a basic nature formed by bacteria. Many of these ptomaïnes when introduced into the body give rise to the same set of symptoms as the bacteria themselves do, so that we may say, *bacteria affect the animal body chiefly through certain toxic principles which they produce and which can be isolated.*

Anti-toxins, Toxins, and Toxalbumens. Late researches claim two classes of products for bacteria—the one *toxic*, and destroyed by heat; the other *anti-toxic*, having a direct action upon the tissue and preventing further infection. Then *proteids* or *toxalbumens*, products extracted from pure cultures, which, like ptomaïnes, produce symptoms similar to those of the bacteria itself. They are amorphous and have no basic action, giving, however, all the reactions of a proteid or albumen.

Filtration of Cultures. These products are isolated from the culture after the bacteria themselves have been separated.

A filter consisting of a cylinder of porcelain, asbestos or kaolin, through which the culture fluid passes, the bacteria remaining behind, is called the Pasteur-Chamberland filter.

Fig. 41.

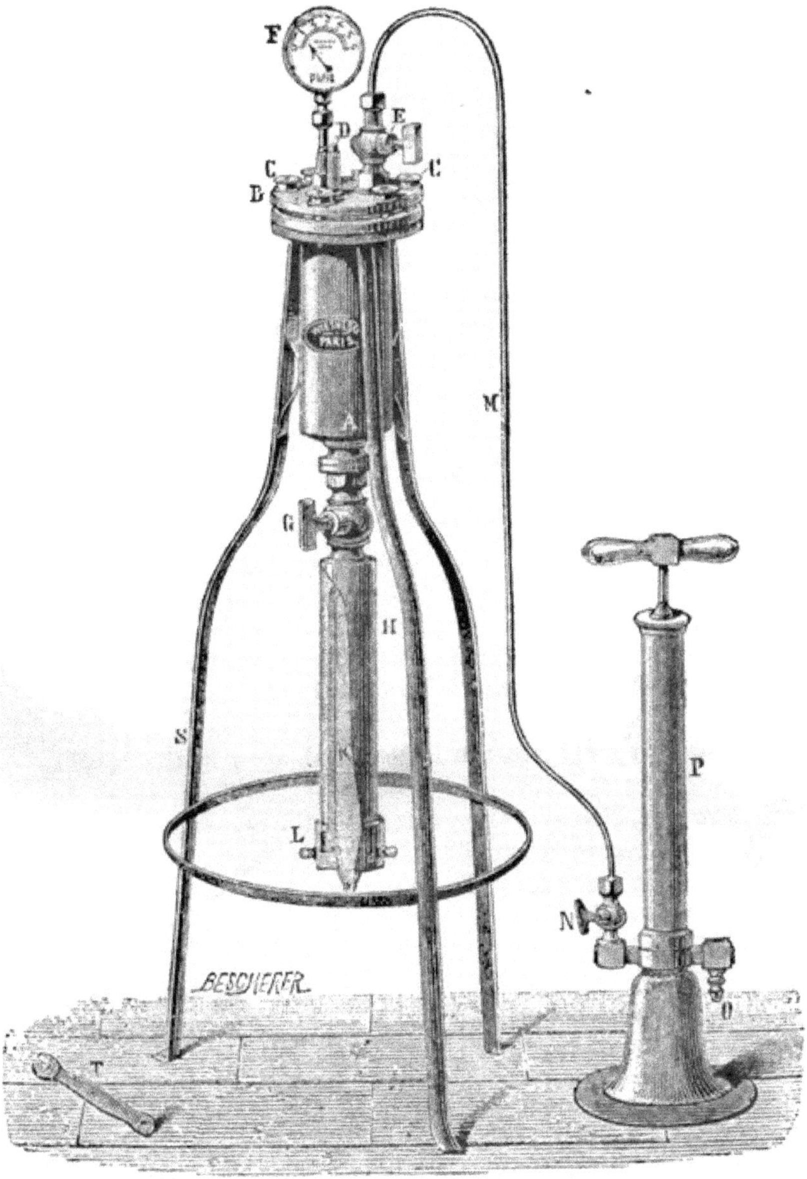

Pasteur-Chamberland filter with pressure.
A. Container. *H.* Filter. *K.* Porcelain. *P.* Air-pump.

(Fig. 41.) The culture can be forced through or *allowed to filter slowly.*

The germless liquor is then treated with various agents, alcohol and acetic acid being that used for the toxalbumen of diphtheria, and a white amorphous powder is at length obtained. These agents have different effects in different doses, and are used also to establish an immunity.

Anti-toxins are obtained by filtering through a Chamberland filter the serum of animals made artificially immune.

They cause immunity when injected into other animals, or cure the already developed disease. The anti-toxins of pneumonia, tetanus, diphtheria, and erysipelas of swine have been *isolated.*

Toxic Bacteria. Those bacteria which produce toxic agents outside of the body, and will not develop in the body, are called *toxic bacteria.* They are pathogenic only in the sense that their products, when accidentally introduced into the body, cause mischief.

Infectious Microbes. Those bacteria which can develop and do develop in the animal body, and there, generate products injurious to the same, are called *infectious bacteria*, or *pathogenic bacteria.*

The Variations of Pathogenesis. The same animals under different circumstances can be differently affected by the same germ.

The ordinary white mouse is not acted upon by the bacillus of glanders.

If, however, glycosuria be produced in the mouse in any way, it speedily becomes attacked by the bacillus.

Different animals are differently affected by the same germs. Ordinarily the white mouse is not acted upon by the bacillus of glanders, but the house mouse is at all times.

The bacterium may first become active when mixed with certain chemicals, it having been harmless before.

Attenuation or Weakened Virulence. Bacteria can be lessened in action either temporarily or permanently, or made inactive entirely without destroying them. There are the natural decay and loss of strength; and successive cultivation in artificial media for a long time of the same germ also destroys its potency.

But artificial means can be used, such as the use of chemical agents added to the nutrient soil, or by passing the germ through animals who are in some sense immune, and are less affected than the animals for whom it is strictly pathogenic.

Thus the bacillus of swine-erysipelas, which is quite virulent for pigs, when passed through rabbits loses much of its power, and again introduced into pigs will sicken them but slightly. Sunlight or any other agent that is destructive to germs will also weaken them when used cautiously.

Heat is the surest agent to lessen the action.

The longer it takes to produce the attenuation, the more lasting it is.

The grade of virulence will oftentimes remain through successive generations.

Some of the attenuations have been named according to the animal that they will affect; thus, *Mice-anthrax* is a culture of anthrax which has been exposed to a temperature of 42.6 C. for twenty days, and which will destroy nothing larger than mice. A culture exposed for ten days will kill nothing larger than rabbits, etc.

The only explanation that can be given of attenuation is that the microbes, though similar in appearance, differ, in that the weaker ones give rise to less toxic products; they have been exhausted.

Nägeli makes use of the simili of the sweet and bitter almond, the one poisonous, because it contains amygdalin, but both possible to be borne on the same tree, and looking alike in every particular.

The Resistance of the Animal Organism to Bacteria. The body is in some sense resistant to bacteria; to some more, to others less; and this resistance has been variously explained.

Chemical Theory. The greater or less alkalinity of the blood diminishing or increasing the virulence is the explanation of some.

The Theory of the Action of the Serum of the Blood. It has been lately shown that the serum of the blood has a direct inhibiting action on all bacteria; and this is directly dependent upon the quantity and quality of albumen in it. It was formerly thought the salts of the blood were the main factors, but these only serve to keep the albumen in good condition.

The serum will hinder the growth of germs, and when bacteria are injected directly into the blood, they soon disappear. They cannot osmose; they collect in the liver, spleen, and marrow of bones, and the corpuscles aid growth. The serum of different animals acts differently upon the same microbe.

Cellular Theory of Metschinkoff. The phagocytes, as he terms them, are anti-microbic. They are the soldiers which endeavor to destroy the enemy. If the cells are strong, they become the victors; if the bacteria are stronger, the bacteria conquer and eat up the cells.

But this theory, though having many supporters, is opposed on sufficient grounds, the one reason being that whenever cells become the residence of live bacteria they suffer; and if Metschinkoff and others have seen bacteria directly enter cells and disappear, it is that they were destroyed before, and that the leucocytes only acted as scavengers. (Fig. 42.)

FIG. 42.

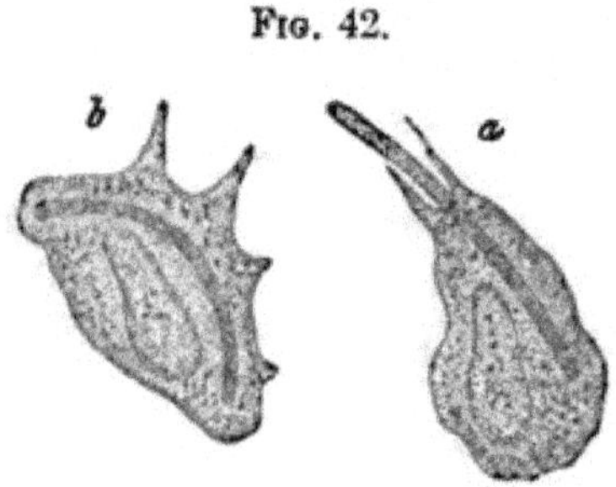

Phagocytes of Metschinkoff: *a.* Bacillus entering; *b.* Bacillus inclosed.

And the late researches with the serum of the blood freed of its cellular elements, and being directly anti-bacteric, would seem to place the phagocytes in the background.

The matter is, however, by no means settled.

To sum up, we have pathogenic microbes such as give rise to products injurious to the animal organism. The infectious ones can overcome the natural resistance of the animal body and develop therein. The bacteria can have this activity lessened or destroyed by agents which are injurious to their products, so as to render them inactive.

CHAPTER XIV.

IMMUNITY.

THE natural or acquired power of resistance to bacterial influences is called immunity.

Natural. As we have mentioned before, certain animals are *naturally* not acted upon by bacteria that affect other animals. We say the animal or person is *immune* by nature.

Acquired. But immunity can be acquired by *various means*. We know that one case of smallpox usually protects against other attacks and so with morbilla and scarlatina. This is through *disease*.

Acclimatization Immunity. Various diseases, which strangers to a climate become affected with, do not trouble the natives.

Artificial Immunity. By the *attenuated virus*, as with anthrax.

Inoculations with *sterilized* cultures, the germs being destroyed. Even certain chemical substances when injected give immunity from certain germs, and when albumen prepared in certain ways was injected, immunity also obtained.

Inoculations intravenously with greatly diluted virulent cultures give immunity.—The various theories which have been made to explain the phenomena of immunity are all unsatisfactory. Some say tnat a first attack destroys the agents which are necessary for the disease to arise. Others say that certain bacterial products remain in the body, and prevent a return of the diseases—act as guards. Some recently claim that the soil is rendered unfit for the further development of the bacteria, after injection of some of the active principles of the bacterium. Some place it in the blood of the animals, it exerting a direct germicidal action.

Alexins. The latest researches place the power of immunity in a certain principle called alexine (αλεξειν, to defend), which has its residence in the serum of the blood and tissues. This principle destroys the bacilli, and they in dying, give out proteids, causing the leucocytes to wander towards these proteids and digest the dying bacilli. The alexins are supposed to be produced by cells containing eosinephile granules.

Chemotaxis. This attraction of leucocytes towards bacteria by means of their chemical products is called chemotaxis.

These proteids when injected by themselves, *i. e.*, sterilized cultures, also produce chemotaxis. When the alexine is not strong enough to destroy the bacterium it develops and finally destroys the action of leucocytes.

Toxines are repellant, producing a "negative chemotaxis."

Cure of Infectious Diseases with Bacteria and their Products. *Antagonism.* It has been known, and is easily demonstrated, that the growth of one bacterium near another results often in the destruction of one, a *direct antagonism* existing.

Rabbits suffering with anthrax were injected with large quantities of streptococcus of erysipelas (pyogenes), and a cure effected, those not so treated dying.

Several other diseases have been so treated in animals with interesting results.

Toxalbumen Injections. When diphtheria is produced in animals, an injection of the toxalbumen of diphtheria will cure the same, and if injected first, diphtheria will not arise. *Tetanus* has been cured and prevented in a similar manner.

Tuberculin. Dixon, in 1889, found, by injecting products of tubercle cultures in glycerine, that Guinea-pigs so treated, suffering from tuberculosis, were cured ; control animals dying. Koch, in 1890, applied this method to man, but without success.

Koch's Rules in Regard to Bacterial Cause of Disease. Before a microbe can be said to be the cause of a disease, it must—

First. Be found in the tissue or secretions of the animal suffering from, or dead with the disease.

Second. It must be cultivated outside of the body on artificial media.

Third. A culture so obtained must produce the disease in question when it is introduced into the body of a healthy animal.

Fourth. The same germ must then again be found in the animal so inoculated.

CHAPTER XV.

EXPERIMENTS UPON ANIMALS.

THE smaller rodents and birds are the ones usually employed for inoculation, as rabbits, Guinea-pigs, rats and mice, and pigeons, and chickens; sometimes monkeys. These are preferred, because easily acted upon by the various bacteria, readily obtained, and not expensive.

The white mouse is very prolific and easily kept, and is therefore a favorite animal for experiment. It lives well upon a little moistened bread. A small box, perforated with holes, is filled partly with sawdust, and in this ten to twelve mice can be kept. When the female becomes pregnant she should be removed to a glass jar until the young have opened their eyes, because the males, which have not been *raised together*, are apt to attack each other.

Guinea-pigs. When Guinea-pigs have plenty of light and air they multiply rapidly. Therefore it is best to have them in some large stall or inclosure. They can be fed upon all sorts of vegetables and grasses, and require but little attention.

Methods of Inoculation. *I. Inhalation.*—Imitating the natural infection, either by loading an atmosphere with the germs in question or by administering them with a spray.

II. Through Skin or Mucous Membrane.

U. With the Food.

Method of Cutaneous Inoculation. The ear of mice is best suited for this procedure. A small abrasion made with the point of a lancet or needle, which has been dipped in the virus. The animal is then separated from the rest and placed in a glass jar, which is partly filled with sawdust and covered with a piece of wire-gauze.

Subcutaneous. The root of the tail of mice is used for this purpose. The hair around the root of the tail is clipped off, and with a pair of scissors a very small pocket is made in the subcutaneous connective tissue, not wounding the animal any more than absolutely necessary, avoiding much blood. The material is placed upon a platinum needle and introduced into the pocket, *solid bodies*, with a forceps.

To hold the mouse still while the operation is going on a little cone made of metal is used. The mouse just fits in here.

There is a slit along the top in which the tail can be fastened, and thus the animal is secure and immobile.

Intravenous Injections. Rabbits are very easily injected through the veins. Mice are too small.

The ear of the rabbit is usually taken. It is first washed with 1-2000 bichloride, which not only disinfects, but also makes the vessels appear more distinct. The base of the ear is compressed to swell the veins. Then a syringe, like the one used for the injection of "tuberculine," a Koch syringe, which can be easily sterilized, is filled with the desired amount of virus and slowly injected into any one of the more prominent veins present. (Fig. 43.)

Intra-peritoneal Injection. This is used with Guinea-pigs mostly. The abdominal wall is pinched up through its entire thickness, and the needle of the syringe thrust directly through, so that it appears on the other side, then the fold let go, the needle withdrawn just far enough so as to be within the cavity.

Inoculation in the Eye. The anterior chamber and the cornea are the two places used. The rabbit is fixed upon a board; the eyelids held apart and head held still by an assistant. A small cut is made in the cornea, a few drops of cocaine having

Fig. 43.

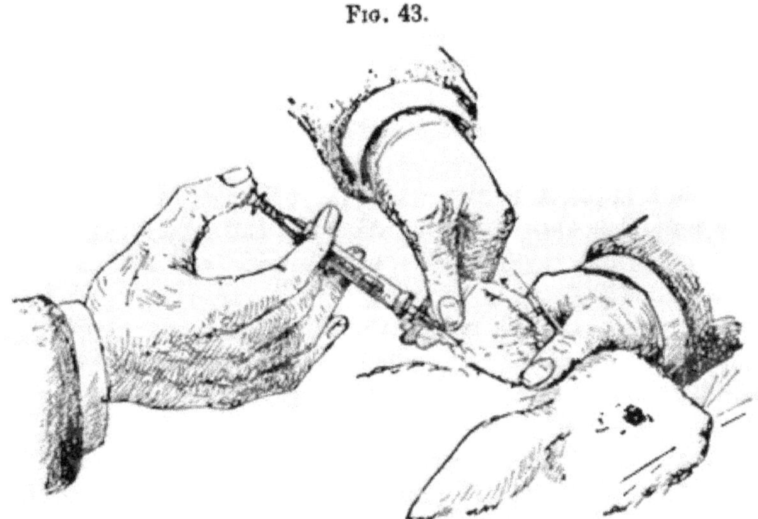

Manner of making intravenous injections in the rabbit.

first been introduced in the eye. The material is passed through the opening with a small forceps, and with a few strokes of a spoon it is pushed in the anterior chamber.

For the cornea a few scratches made in the corneal tissue will suffice ; the material is then gently rubbed in.

Inoculation of the Cerebral Membranes. The skin and aponeurosis cut through where the skull is the thinnest. Then the bone carefully trephined, and the dura exposed. In *Rabies* inoculation, the syringe containing the hydrophobic virus pierces the dura and arachnoid, and the virus is discharged beneath the latter.

Intra-Tracheal. The bacteria can be introduced directly into the trachea, thus coming in contact with the lungs.

Intra-duodenal.—Cholera germs are injected into the intestines after they have been exposed, by carefully opening the abdomen. This is done in order to avoid the action of the gastric juice.

Obtaining Material from Infected Animals. The animal should be skinned, or the hairs plucked out, before it is washed, at least the portion where the incision is to be made. Then the entire body is washed in sublimate. Two sets of instruments are required, one for coarser and one for finer work : the one sterilized in the flame ; the other, to prevent being damaged, heated in a hot air oven.

The animal, the mouse for example, is stretched upon a board, a nail or pin through each leg, and the head fixed with a pin through the nose. The skin is dissected away from the belly without exposing the intestines. Then the ribs being laid bare, the sternum is lifted up, and the pericardium exposed. A platinum needle dipped into the heart after the pericardium has been slit will give sufficient material for starting a culture. If the other organs are to be examined, further dissection is made. If the intestines were first to be looked at, they would be laid bare first.

In this manner material is obtained, and the results of inoculation noted.

Frequent sterilization of the instruments is desirable.

PART II.

SPECIAL BACTERIOLOGY.

CHAPTER I.

NON-PATHOGENIC BACTERIA.

Special Bacteriology. Under this head the chief characteristics of individual bacteria will be detailed, *pathogenic* and *non-pathogenic* being the main divisions. It is usual to describe the non-pathogenic first.

Non-Pathogenic Bacteria. We can give but a few of the more important varieties.

Bacillus Prodigiosus. (Ehrenberg.) This bacillus, formerly called a micrococcus, is very common, and one of the first noticed, because of the lively red color it forms on vegetables and starchy substances. "The bleeding host," miracles being due to it.

Form.—Short rods, often in filaments, *without spores.*

Immobile.—Has no automatic *movements.*

Facultative anærobic, that is, it can grow without air; but the pigment requires oxygen to show itself.

Growth. Gelatine. Liquefy rapidly.

Colonies.—At first white, round points with smooth edge appearing brown under microscope, but soon changing to red.

Stab Cultures.—The pigment develops on the surface, the growth occurring all along the line.

Potato is well suited to the growth, the pigment developing after twelve hours. *Agar* and *blood serum* growths do well.

Temperature.—Grows best at 25° C.

Varieties.—By exposure to heat of brood-oven during several generations the power to produce pigment can be temporarily abolished.

The Pigment.—A pigment-forming body is created by the bacillus, and the action of oxygen upon it produces the color. It is insoluble in water, slightly soluble in alcohol and ether; acids fade it, alkalies restore the color.

Gases.—A trimethylamin odor arises from all cultures.

Stain.—Takes all *aniline dyes easily* in the ordinary way.

Bacillus Indicus. (Koch.)

Syn. Micrococcus Indicus. Origin.—Found in the stomach of an Indian ape.

Form.—Short rods with rounded ends. No *spores. Automatic movements* present; *facultative anærobin.*

Growth. Gelatine.—Liquefy rapidly.

Colonies.—Round, or oval, granular margins; brilliant *red pigment.*

Stab Cultures.—On the surface the pigment shows itself. Grows well on other media.

Temperature.—Grows best at 35° C.

Action on Animals.—In very large quantities, if injected into the blood, a severe and fatal gastro-enteritis can be produced.

Stain.—Takes all dyes.

FIG. 44.

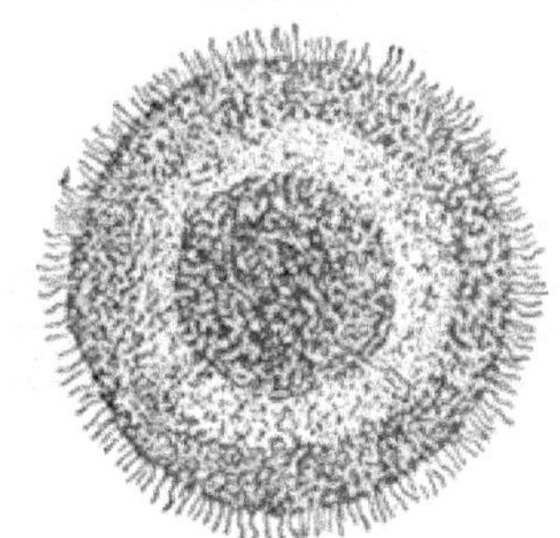

Colony of Bacillus Mesentericus Vulgatus.

Bacillus Mesentericus Vulgatus.

The common potato bacillus of *Flügge.*

Habitat.—Surface of the soil, on potatoes, and in milk.

Form.—Small thick rods with rounded ends, often in pairs.

Properties.—Very motile; produce abundant spores; liquefy gelatine; diastatic action.

Growth.—Rapid.

Plate Colonies.—Round, with transparent centre at first, then becoming opaque. The border is ciliated; little projections evenly arranged.

Potato.—A white covering at first, which then changes to a rough brown skin; the skin can be detached in long threads.

Temperature.—Spores at ordinary temperatures.

Spores.—Are very resistant; are colored in the manner described in first part of the book for spores in general.

Bacillus Megaterium (de Bary).

Origin.—Found on cooked cabbage.

Form.—Large rods, four times as long as they are broad, 2.5 μ. Thick rounded ends. Chains with ten or more members often formed granular cell contents.

Properties.—Abundant spore formation; very slow movement; slowly dissolves gelatine.

Growth.—Strongly ærobic; grows quickly, and best, at a temperature of 20° C.

Plate Colonies.—Small, round, yellow points in the depth of the gelatine. Under microscope irregular masses.

Fig. 45.

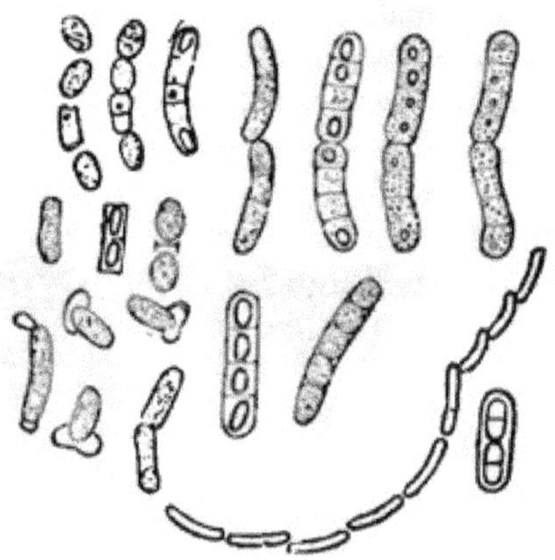

Bacillus Megaterium, with spores.

Stab Culture.—Funnel-shaped from above downwards.

Potato.—Thick growths, with abundance of spores.

Bacillus Ramosus.

Syn. Bac. Mycoides (Flügge). *Wurzel* or *root bacillus.*

Origin.—In the upper layers of garden or farm grounds, and in water.

Form.—Short rods, with rounded ends, about three times as long as they are thick; often in long threads and chains

Properties.—Large, shining, oval spores; a slight movement; *liquefy gelatine.*

Growth.—At ordinary temperatures, with plentiful supply of air.

Plate Colonies.—Look like roots of an old tree gnarled together, radiating from a common centre. On surface soon liquid.

Stab Culture.—Soon a growth occurs along the needle track, and the whole resembles a pine tree turned upside down. The gelatine then becomes liquid, a thin skin floating on top, and small flakes lying at the bottom.

Stroke Culture.—Feathery resemblance is produced.

Staining.—Spores stain readily with the ordinary spore stain.

Bacterium Zopfi. (Kurth.)

Origin.—Intestines of a fowl.

Form.—Short thick rods forming long threads coiled up, which finally break up into spores, which were once thought to be micrococci.

Properties.—Very motile; does not dissolve or liquefy gelatine.

Growth.—In thirty hours abundant growth; *ærobic;* grows best at 20° C.

Plates.—Small white points which form the centre of a very fine netting. With high power this netting is found composed of bacilli in coils, like braids of hair.

Excellent impress or "Klatsch" preparations are obtained from these colonies.

Staining.—Ordinary dyes.

Bacillus Subtilis. (Hay Bacillus.) Ehrenberg.

Origin.—Hay infusions; found also in air, water, soil, fæces, and putrefying liquids.

Form.—Large rods, three times as long as broad; slight roundness of ends, transparent; seldom found singly; usually in long threads. *Flagella* are found on the ends. *Spores of* oval shape, strongly shining, very resistant.

Properties.—Very motile; dissolves gelatine.

Growth.—Rapid; strongly ærobic.

Plate.—Round, gray colonies, with depressed white centre.

Under microscope the centre yellow; the periphery like a wreath, with tiny little rays projecting; very characteristic.

Potato.—A thick moist skin forms in twenty-four hours.

Staining.—Rods, ordinary stain, *spores*, spore stain.

It is easily obtained by covering finely cut hay with distilled water, and boiling a quarter of an hour. Set aside forty-eight hours. A thick scum will show itself on the surface composed of the subtilis bacilli.

Bacillus Spinosus. (Lüderitz.)

Called spinosus because small spine-like processes are formed by the colonies.

Origin.—In the juices of the body of a mouse and guinea-pig which were inoculated with garden earth.

Form.—Large rods, straight, some slightly bent, ends rounded; often in long threads.

Properties.—Large spores, the bacillus enlarging to allow the spores to develop; very motile; gelatine slowly liquefied. A gas is formed in the culture having an odor like Swiss cheese.

Growth.—The growth occurs at ordinary temperatures only when the oxygen is *excluded.* Very strongly anærobic. Glucose added to the gelatine (1 to 2 per cent.) increases the nutritive value.

Colonies in *roll cultures* and *high stab cultures* appear as little spheres surrounded by a zone of liquefied gelatine. In the deeper growths thorn-like projections or spines develop proceeding from a gray-colored centre.

Staining.—With ordinary methods. This bacillus, being strongly anærobic, must be cultured with the usual care taken with anærobins.

Some Bacteria found in Milk. Bacillus Acidi Lactici. (Hüppe.)

Origin.—In sour milk.

Form.—Short thick rods, nearly as broad as they are long, usually in pairs.

Properties.—Immotile. Spores large shining ones. Do not liquefy gelatine. Breaks up the sugar of milk into lactic acid and carbonic acid gas, the casein being thereby precipitated.

Growth.—Slow; is facultative anærobic. Grows first at 10° C.

Plate Colonies.—First small white points, which soon look like porcelain, glistening. Under microscope the surface colonies resemble leaves spread out.

Stab Culture.—A thick dry crust with cracks in it forms on the surface after a couple of weeks.

Attenuation.—If cultured through successive generations, they lose the power to produce fermentation. Several other bacteria will give rise to lactic acid fermentation; but this especial one is almost constantly found, and is very wide-spread.

In milk, it first produces acidity, then precipitation of casein, and finally, formation of gases.

A bacillus described by Grotenfeldt, and called *Bacterium Acidi Lactici*, forms alcohol in the milk. It was found in milk in Bavaria.

Bacillus Butyricus. (Hüppe.)

This bacillus causes butyric acid fermentation.

Origin.—Found in milk.

Form.—Short and long thin rods with rounded ends; large oval spores, seldom forming threads.

Properties.—Very motile; liquefies gelatine rapidly; produces gases resembling butyric acid in odor. In milk it coagulates the casein, decomposes it, forming peptones and ammonia, with a bitter taste, and butyric acid fermentation. An alkaline reaction.

Growth.—Quickly, at 35° to 40° C., with oxygen. Spons very resistant.

Colonies. Plate.—Small yellow points which soon run together, becoming indistinguishable.

Stab Culture.—A small yellow skin formed on the surface with delicate wrinkles; cloudy masses in the liquefied portion.

Staining.—With ordinary stains.

Bacillus Amylobacter (Van Tiegham); or, **Clostridium Butyricum.** (Prazmowsky.) (Vibrion butyrique of Pasteur.)

Origin.—Found in putrefying plant-infusions, in fossils, and conifera of the coal period.

Form.—Large, thick rods, with rounded ends, often found in chains. A large glancing spore at one end, the bacillus becoming spindle-shape in order to allow the spore to grow; hence the name clostridium.

Fig. 46.

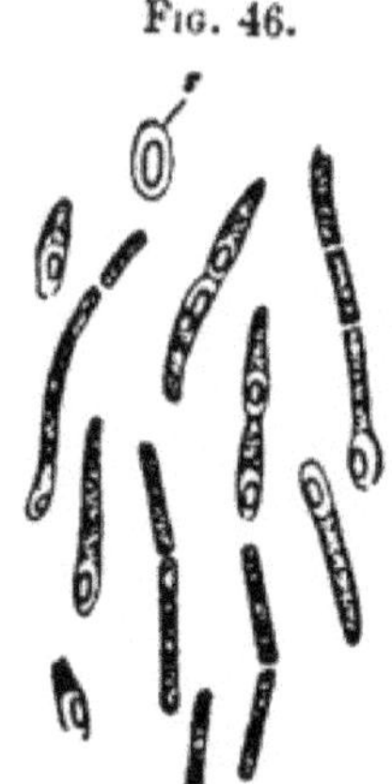

Bacillus Amylobacter.

Properties.—Very motile; gases arise with butyric smell. In solutions of sugars, lactates and cellulose-containing plants, and vegetables, it gives rise to decompositions in which butyric acid is often formed. Casein is also dissolved.

Like granulose, a watery solution of iodine will color blue some portions of the bacillus; therefore it has been called *amylobacter*.

Growth.—It is strongly anærobic, and has not yet been satisfactorily cultivated.

Bacillus Lactis Cyanogenus. *Bacterium Syncyanum,* (Hüppe.)

Origin.—Found in blue milk.

Form.—Small narrow rods about three times longer than they are broad; usually found in pairs. The ends are rounded.

Properties.—They are very motile; do not liquefy gelatine; form spores usually in one end. A bluish-gray pigment is formed outside of the cell, around the medium. The less alkaline the media the deeper the color. It does not act upon the milk otherwise than to color it blue.

Growth.—Grows rapidly, requiring oxygen. *Colonies on plate.* Depressed centre surrounded by ring of porcelain-like bluish growth. Dark brown appearance under microscope.

Stab Culture.—Grows mainly on surface; a nail-like growth. The surrounding gelatine becomes colored brown.

Potato.—The surface covered with a dirty blue scum.

Attenuation.—After prolonged artificial cultivation loses the power to produce pigment.

Staining.—By ordinary methods.

Bacillus Lactis Erythrogenes. *Bacillus of Red Milk.* (Hüppe and Grotenfeldt.)

Origin.—Found in red milk, and in the fæces of a child.

Form.—Short rods, often in long filaments, without spores.

Properties.—Does not possess self-movement. Forms a nauseating odor; liquefies gelatine. Produces a yellow pigment which can be seen in the dark, and a *red pigment* in alkaline media, away from the light. In milk it produces the yellow cream on top of the blood-red serum, or, fluid in the centre, and at the bottom the precipitated casein.

Growth.—Grows rapidly in bouillon and on potatoes; slower on the other media; *Plates.* A cup-like depression in the centre of the colony, with a pink coloration around it, the colony itself being slightly yellow.

Stab Culture.—The growth mostly on surface. The gelatine afterwards colored red and liquefied.

Potato.—A golden yellow pigment formed at 37° C., after six days.

Some Non-Pathogenic Bacteria found in Water. The bacteria found here are very often given to producing pigments or phosphorescence, and are in great number. The more common ones only will be described.

Bacillus Violaceus.

Origin.—Water.

Form.—A slender rod with rounded ends, three times as long as it is broad, often in threads; middle-sized spores.

Properties.—Very motile; forms a violet-blue pigment, which is soluble in alcohol, and depends upon oxygen for its growth. Rapidly liquefies gelatine, but not agar.

Growth.—Grows fairly quick, is facultative anærobic.

Cultures on Plate.—At first the colonies look like inclosed air-bubbles. Low power shows irregular masses, with a centre containing the pigment and a hairy-like periphery.

Stab Culture.—Cone-like liquefaction containing air, and the pigment, in separated granules, lying towards the bottom.

Stroke Culture on Agar.—A violet, ink-like covering which remains for years.

Bacillus Cœruleus. (Smith.)

Origin.—Schuylkill water.

Form.—Very thin rods; 2.5 μ. long, 0.5 μ. wide; often in threads; *spores* were not found.

Properties.—Liquefies gelatine; produces a very deep-blue pigment.

Growth.—Slowly, with oxygen, at ordinary temperature.

Plate.—Round colonies on the surface of bluish color.

Stab Cultures.—A cup-shaped liquefaction along the needle thrust, with a sparse growth, the liquefied portion appearing blue.

Fluorescent Bacteria. Several kinds present in water.

Bacillus Erythrosporus. (Eidam.)

Origin.—Drinking water and putrefying albuminous solutions.

Form.—Slender rods often in short threads, with spores of oval shape, and appearing as if stained with fuchsin.

Properties.—Motile; does not dissolve gelatine; produces a greenish-fluorescent pigment which appears yellow in reflected light, but green on transmitted light.

Growth.—Somewhat quickly; facultative anærobic; growth only at ordinary temperatures.

Plates.—White colonies, with greenish-yellow fluorescence around each colony. Under microscope the periphery appears radiated.

Stab Cultures.—Good growth along the needle thrust; the whole gelatine gives out the fluorescence.

Bacillus Fluorescens Liquefaciens.

Origin.—Water, and from conjunctival sac.

Form.—Very fine little rods; no spores.

Properties.—Motile; forms a greenish-yellow fluorescent pigment; liquefies gelatine.

Growth.—Rapid only at ordinary temperatures, and strongly ærobic.

Plates.—Round colonies, cup-shaped depressions, the solid gelatine that remains becoming colored with greenish-yellow fluorescence.

Stab Culture.—On the surface, air-bubble depressions; the white colonies in the bottom of these depressions, and the solid gelatine around the inoculation shining with the fluorescence.

Phosphorescent Bacteria. Six varieties of phosphorescent bacteria have been described; they are found usually in sea-water, or upon objects living in the sea.

Bacillus Phosphorescens Indicus. (Fischer.)

Origin.—Tropical waters.

Form.—Thick rods, with rounded ends, sometimes forming long threads.

Properties.—Very motile; liquefying gelatine at a temperature of 25° to 30° C., with oxygen and a little moisture, and in the dark, a peculiar electric-blue light develops a phosphorescence.

Growth.—Slowly; must have oxygen; does not grow under 10° C. or over 50° C.

Plates.—Little round, gray points, which under low power appear as green colonies with reddish tinge around them. *Cooked fish*, when smeared upon the surface with a little of the culture, show the phosphorescence most marked. Grows well on *potatoes* and *blood-serum.*

Bacillus Phosphorescens Indigenus. (Fischer.)

Origin.—Waters in the northern part of Germany. It differs from the Indian bacillus, in that it grows at a temperature of 5° C., and does not develop upon potatoes or blood-serum.

Bacillus Phosphorescens Gelidus. (Förster.)

Origin.—Surfaces of salt-water fish.

Form.—Short, thick rods, looking oval sometimes; zoogl[oe]a are often formed.

Properties.—Motile; does not liquefy gelatine; a beautiful phosphorescence from the surface of fish; it can be photographed by its own light.

Colonies.—Grows best between 0° and 20° C.; grows slowly, and mostly on the surface. The material must contain salt. A bouillon made with sea-water, or 3 to 4 per cent. common salt will suffice. The colonies appear as those of the Phosphorescens Indicus.

Fresh herring laid between two plates will often show phosphorescence in twenty-four hours.

The other three varieties require *glucose* in the culture before they give out any glow. They are *Bacterium Pflugeri*, *Bact. Fischeri*, and *Bact. Balticum.* They do not dissolve gelatine.

Several very indistinct species, found in waters from factories and in some of the mineral waters, deserve yet to be mentioned. They have been given various names by observers; almost a new classification created. Such are the *crenothrix*, *cladothrix*, and *beggiatoa.*

Crenothrix Kühniana. (Rabenhorst.) Long filaments joined at one end; little rod-like bodies form in the filaments; and these break up into spores.

Zoogloea are also formed by means of spores; and these can become so thick as to plug up pipes and carriers of water. They are not injurious to health.

Cladothrix Dichotoma. (Cohn.) Very common in dirty waters. The filaments branch out at acute angles, otherwise resembling the crenothrix; accumulations of ochre-colored slime, consisting of filaments of this organism, are found in springs and streams.

Beggiatoa Alba. (Vancher.) The most common of this species. The distinction between this and the preceding species lies in the presence of sulphur granules contained in the structure, and hence they are often found where sulphur or sulphides exist; but where the remains of organic life are decomposing they can also be found.

Several large spirilla and vibrio live in bog and rain-water, but our space does not suffice to describe them.

Bacterium Ureæ.

Origin.—Decomposed ammoniacal urine.

Form.—Thick, little rods, with round ends one-half as thick as they are long.

Properties.—Does not dissolve gelatine; changes urea into carbonate of ammonia.

Growth.—At ordinary temperatures, very slowly. In two days on gelatine very minute points, which in ten days have the size of a cent. The colonies grow in concentric layers.

Micrococcus Ureæ. (Pasteur and Van Tieghain.)

Origin.—Decomposed urine and in the air.

Form.—Cocci, diplococci, and steptococci.

Properties.—Decomposes urea into carbonate of ammonia; does not liquefy gelatine.

Growth.—Grows rapidly, needing oxygen; can remain stationary below 0° C.; growing again, when a higher temperature is reached.

Colonies on Plate.—On the surface like a drop of wax.

Stab Cultures.—Looks like a very delicate thread along the needle thrust.

Spirillum. Spirillum Rubrum. (Esmarch.)

Origin.—Body of a mouse dead with septicæmia.

Form.—Spirals of variable length, long joints, flagella on each end ; no spores.

Properties.—Does not liquefy gelatine ; very motile ; produces a wine-red pigment, which develops only by absence of oxygen.

Growth.—Can grow with oxygen, but is then colorless ; grows very slowly ; ten to twelve days before any sign ; grows best at 37° C.

Gelatine Roll Cultures.—Small, round ; first gray, then wine-red colonies.

Stab Cultures.—A red-colored growth along the whole line ; it is deepest below, getting paler as it approaches the surface.

Spirillum Concentricum. (Kitasato.)

Origin.—Decomposed blood.

Form.—Short spirals, two to three turns, with pointed ends ; it has flagella on the ends.

Properties.—Very motile ; does not liquefy gelatine.

Growth.—Very slow ; mostly on the surface ; best at ordinary temperatures.

Plates.—A growth of rings concentrically arranged, every alternate one being transparent ; the furthest one from the centre possessing small projections.

Stab Cultures.—Growth mostly on the surface.

Sarcina. *Cocci* in cubes or packets of colonies. A great number have been isolated ; many producing very beautiful pigments. The majority of them found in the air.

Sarcina Lutea. (Schröter.)

Origin.—Air.

Form.—Very large cocci in pairs ; tetrads and groups of tetrads.

Properties.—Liquefies gelatine slowly ; produces sulphur-yellow pigment.

Growth.—Slowly ; at various temperatures ; strongly ærobic.

Plates.—Small, round, yellow colonies.

Stab Cultures.—Grows more rapidly, the growth being nearly all on the surface, a few separated colonies following the needle thrust for a short distance. *Agar*, a very beautiful yellow, along the stroked surface.

Sarcina Aurantica.—*Flava*, *roseo*, and *alba* are some of the other varieties. Many are *obtained* from beer.

Sarcina Ventriculi. (Goodsir.)

Origin.—Stomach of man and animals.

Form.—Colorless, oval cocci, in groups of eight and packets of eight.

Properties.—Does not liquefy gelatine; shows the reaction of *cellulose* to *iodine*.

Growth.—Rapid. At end of thirty-six hours, round, yellow colonies, from which colorless cocci and cubes are obtained.

Habitat.—They are found in many diseases of the stomach, especially when dilatation exists.

CHAPTER II.

PATHOGENIC BACTERIA.

We have divided this part into two portions.

I. Those bacteria which are pathogenic for man and other animals.

II. Those bacteria which do not affect man, but are pathogenic for the lower animals.

Here again it will only be possible to give the more important bacteria; there are many diseases in which micro-organisms have been found, but they have not yet been proven as causative of the disease, and have also been found in other diseases. We cannot treat of them here.

Bacillus Anthracis. (*Rayer and Davaine.*)—Rayer and Davaine, in 1850, first described this bacillus; but *Pasteur*, and later *Koch*, first gave it the importance it now has.

Synonyms.—Bacteridie du charbon (Fr.), Milzbrand bacillus (German); bacillus of splenic fever, or malignant pustule.

Origin.—In blood of anthrax-suffering animals.

Form.—Rods of variable length, nearly the size of a human blood-corpuscle, broad cup-shaped ends; in bouillon cultures, long threads are formed, with *large oval spores.*

Fig. 47.

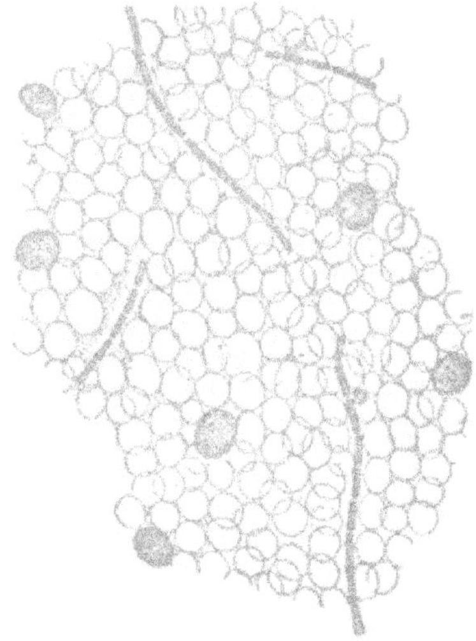

Anthrax bacilli in human blood (fuchsin staining), Zeiss 1-12 oil immersion. No. 4 ocular taken from Vierordt.

Properties.—Liquefies gelatine; immotile; the spores are very resisting, living twenty years.

Growth.—Grows rapidly, between 12° C. and 45° C., and requires plenty of oxygen, not growing without it; grows well in all media.

Plates of Gelatine.—Colonies develop in two days, white shiny spots, which appear under microscope as slightly yellowish granular-twisted balls, like a ball of yarn; each separate string or hair, if looked at under high power, being composed of bacteria in line.

Stab Cultures.—A white growth with thorn-like processes along the needle-track; later on, gelatine liquefied, and flaky masses at the bottom.

Potato.—A dry creamy layer, and when placed in brood-oven, rich in spores.

Varieties. Asporogenic.—By cultivation in gelatine, containing 1 to 1000 ac-carbolic, a variety develop that cannot produce spores. Also *involution forms*, differing from the usual type.

Fig. 48. Fig. 49.

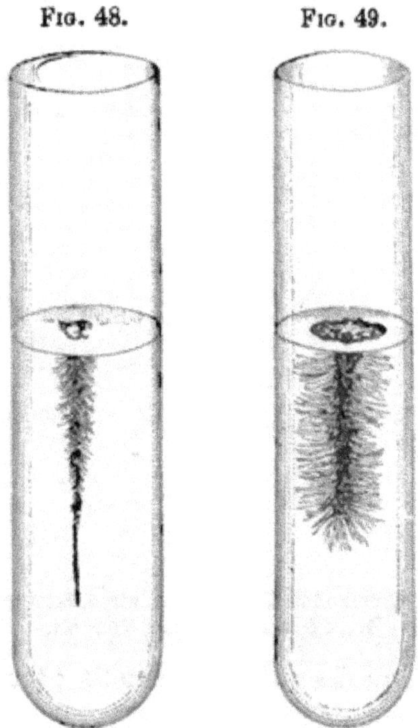

Stab Cultures of Anthrax in Gelatine.

Staining.—They readily take all the aniline dyes with the ordinary methods. To bring out the cup-shaped concave extremities, a very weak watery solution of methylin blue is best.

Spores are stained by the usual method. When several bacilli are joined together, the place of their joining looks like a spore because of the hollowed ends. The double staining will develop the difference.

Sections of tissue are stained according to the ordinary methods, taking *Gram's* method very *nicely*.

Pathogenesis.—When mice are inoculated with anthrax material through a wound in the skin, they die in twenty-four hours from an active septicæmia, the point of inoculation remaining unchanged. The following appearances then present themselves :—

Peritoneum.—Covered with a gelatinous exudate.

Spleen.—Very much swollen, dark red, and friable.

Liver.—Parenchymatous degeneration.

Blood.—Dark red. The bacilli are found wherever the capillaries are spread out, in the spleen, liver, intestinal villi, and glomeruli of kidney, and in the blood itself. Only when the capillaries burst are they found in the tubules of the kidney.

Mode of Entrance.—The bacilli can be *inhaled*, and then a pneumonia is caused, the pulmonary cells containing the bacilli ; when the spores are inhaled, a general infection occurs.

Feeding.—The cattle graze upon the meadows, where the blood of anthrax animals has flowed and become dried, the spores remaining, which then mix with the grass and so enter the alimentary tract ; here they then cause the intestinal form of the disease, ulcerating through the villi.

Local Infection.—In man usually only a local action occurs ; by reason of his occupation—wool-sorter, cattle-driver, etc., he obtains a small wound on the hand, and local gangrene and necrosis set in.

Pneumonia by inhalation can also occur in man.

Susceptibility of Animals.—Dogs, birds, and cold-blooded animals affected the least ; while mice, sheep, and guinea-pigs quickly and surely.

Products of Anthrax Bacilli.—A basic ptomaïne has not been found, but a toxalbumen or proteid, called *anthraxin*, has been obtained. A certain amount of *acid* is produced by the virulent form, *alkali* by the weak.

Attenuation and Immunity.—Cultures left several days at a temperature between 40° and 42° C. soon become innocuous, and when injected into animals protect them against the virulent form.

The lymph obtained from lymph-sac of a frog destroys the virulence of anthrax bacilli and spores temporarily.

Hankin obtained an alexin from the blood and spleen of rats, they being naturally immune. It destroyed the anthrax bacilli in vitro, and used by injection in susceptible animals made them immune. It is insoluble in alcohol or water.

Protective Vaccination.—Animals have been rendered immune by various ways—by inoculation of successive attenuated cultures; also with sterilized cultures—that is, cultures containing no bacilli, and with cultures of other bacteria.

Habitat.—The anthrax disease seems confined to certain districts in Siberia, Bavaria, and Auvergne, and mainly during the summer months.

The bacillus has never been found free in nature.

Bacillus Tuberculosis. (Koch.)

This very important bacillus was first described, demonstrated, and cultivated by Koch, who made his investigations public on the 24th of March, before the Physiological Society of Berlin, in the year 1882.

FIG. 50.

Tubercle bacilli in sputum, carbol-fuchsin, and methylin blue. Zeiss 1:12 oil immersion.

Origin.—In various tubercular products of man and other animals.

Form.—Very slender rods, nearly straight, about one-quarter the size of a red corpuscle's diameter, their ends rounded, usually solitary, often, however, lying in pairs in such a manner as to form an acute angle. Sometimes they are 'S'-shaped. In colored preparations little oval spaces are seen in the rod, which resemble spores; but the question of the existence of spores is still undecided.

Properties.—Does not possess self-movement.

Growth.—Requires special media for its growth, and a temperature varying but slightly from 37.5° C. It grows slowly, developing first after ten days, reaching its maximum in three weeks. It is facultative anærobic. On gelatine it does not form a growth.

Colonies on Blood Serum.—Koch first used blood serum for culture ground, and obtained thereon very good growths. Test-tubes with *stroke* culture were placed in the brood oven at 37° C. for ten to fourteen days, when small glistening white points appeared which then coalesced to form a dry, white, scale-like growth. Under microscope composed of many fine lines containing the tubercle bacillus.

Glycerine Agar.—By adding four to six per cent. glycerine to ordinary agar-peptone medium, Nocard and Roux obtained a culture ground upon which tubercle bacilli grew much better than upon blood serum. This is now almost exclusively used.

Stroke cultures are here used as with blood serum. They are placed in brood-oven after inoculation, and remain there about ten days, at a temperature of 37° C. The cotton plugs of the tubes are covered with rubber caps, the cotton first having been passed through the flame, and moistened with a few drops of sublimate solution. The rubber cap prevents the evaporation of the water of condensation which always forms, and keeps the culture from drying up.

Fig. 51.

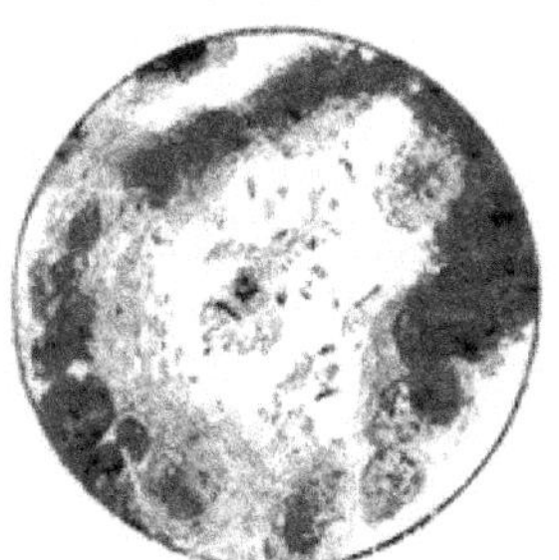

Tubercle bacilli in human liver 500 ×. (Fränkel and Pfeiffer.)

The growth which occurs resembles the rugæ of the stomach, and sometimes looks like crumbs of bread moistened. The impression or "Klatsch" preparation shows under the microscope a thick curled-up centre around which threads are wound in all directions. And these fine lines show the bacilli in profusion.

Fig. 52.

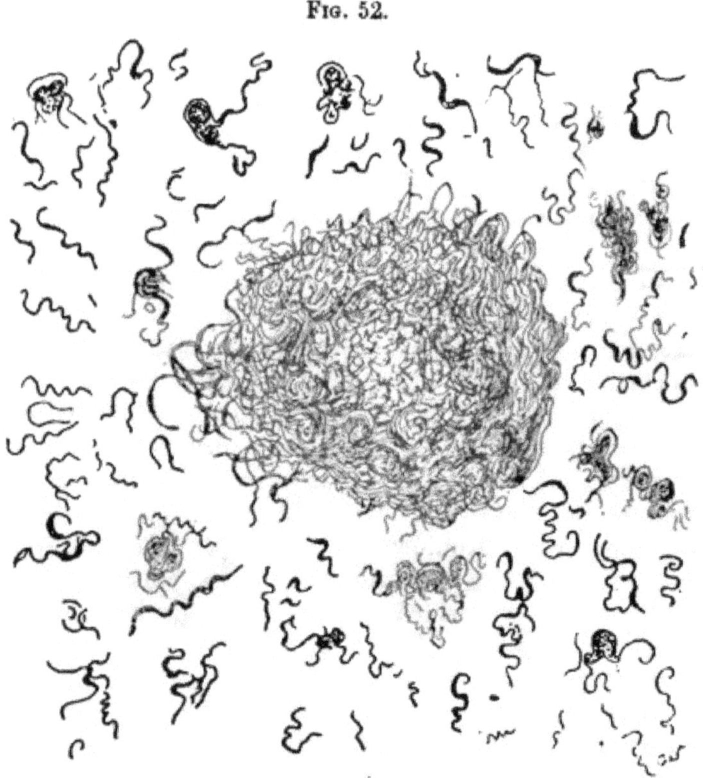

Klatsch preparation.

Potato.—It can be cultivated on slices of potato which are placed in air-tight test-tubes.

Bouillon.—Bouillon containing four per cent. glycerine is a very good nurture ground.

Varieties.—Dixon, of Philadelphia, and others have obtained branched forms of bacilli and club-shaped forms.

In Sputum.—Little granules arranged like streptococci, which take the characteristic stain, and look as if the protoplasma had been destroyed that enclosed them.

Fig. 53.

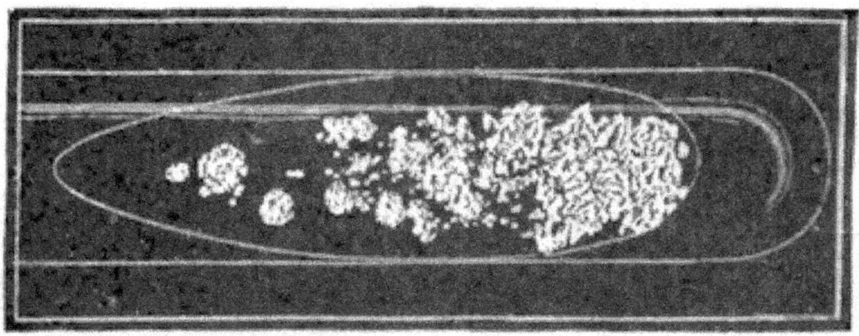

Growth on Agar.

Bovine tubercle-bacilli are about one-third smaller than human tubercle bacilli.

Staining.—The tubercle bacilli require special methods to stain them, and a great number have been introduced. They are stained with great difficulty: but once stained, they are very resistant to decolorizing agents. Upon these facts all the methods are founded.

It will only be necessary to describe those methods principally in use; and as the examination of sputum for bacilli is of so frequent an occurrence, and so necessary, it is well to detail in particular the method of staining.

Starting with the sputum, we search for little clumps or rolled-up masses; if these are not present, the most solid portions of the mucus are brought with forceps upon a clean cover-glass; very little suffices. With another cover-glass it is pressed and spread out evenly; drawing one glass over the other, we obtain two specimens, and these put aside or held high over the flame until dry.

If we desire to examine the specimen quickly, or make a hurried diagnosis, we use the *rapid* method, with hot solutions; otherwise we let it stay, in cold solution until the next morning the advantages of which will be later on described.

The Rapid Method.—(B. Fränkel's method modified by Gabbet.) The principle is to combine with the contrast stain the decolorizing agent; but the preparations are not permanent; the method, however, is very useful.

Two solutions are required: one of Ziehl's carbol-fuchsin; the other Gabbet's acid methylin blue. (See No. X., on page 33.)

The cover-glass containing the dried sputum is passed three times through the flame, as described in the general directions. It is then placed in the carbol-fuchsin solution five minutes (cold), or two minutes in the hot, immediately then transferred to the second solution, the acid blue, where it remains one minute, then washing in water. The preparation is dried between filter-paper, and mounted best first in water. Examined with oil-immersion.

Another Rapid Method.—This method possesses the advantage of giving permanency to the preparations; but the bacilli are distorted and ugly crusts form.

Three dishes are required:—

The first contains nitric acid and water 1–4.

The second alcohol.

The third distilled water.

They are arranged one after the other in the above order.

Two staining solutions must be at hand, carbol-fuchsin and watery methylin blue. The cover-glass containing the dried sputum is passed three times through the flame, then covered with a few drops of carbol-fuchsin, and held in the forceps over the flame, so that the stain will boil upon the glass. With a pipette the dye is constantly added and kept boiling for about one minute. It is then decolorized by holding it in the first dish until it appears brownish-black, then directly into the second dish, when the alcohol peals off the red color in little clouds, and it becomes nearly colorless; about five seconds suffice. Then it is placed in the third dish, the water washing off the alcohol. If the color of the preparation is now deeper than a slight pink, it is again dipped into the acid, alcohol, and back into the water, careful not to hold it too long in the above solutions. The contrast stain is now applied, a few drops of the methylin blue solution left on cold for two minutes being sufficient. The glass is now dried and mounted on the slide in

Canada balsam. Examined with oil-immersion. The tubercle bacilli red, all else blue.

Slow Method.—When perfect permanent preparations are desired and the bacilli to be seen unaltered, the slow method is to be preferred, and it is to be recommended whenever the time allows. It consists simply in allowing the carbol-fuchsin to work upon the preparation a number of hours. We usually place the cover glass with the dried sputum and which has been drawn through the flame three times, in a little dish containing enough dye to allow the glass to be immersed. We do this about 5 or 6 o'clock P.M., and the next morning the preparation is ready for decolorizing, the process being the same as described above, viz., 25% nitric acid, alcohol and water, and the contrast stain methylin blue. We thus avoid the formation of ugly crusts, the bacilli are not distorted, the specimen is permanent and very clear.

Biedert's Method of Collecting Bacilli, when the bacilli are very few in a great quantity of fluid, as urine, pus, abundant mucus, etc., Biedert advises to mix 15 c.cm. of the fluid with 75 to 100 c.cm. water and a few drops of potassium or sodium hydrate, then boiling until the solution is quite thin. It is placed in a conical glass for two days, and bacilli with other morphological elements sink to the bottom of the glass; when the supernatant liquid is decanted, the residue can be easily examined. In this way bacilli were found that had eluded detection examined in the ordinary manner.

The centrifugal machine is used either in connection with Biedert's sediment method or without, to obtain the solids suspended in urine or serum.

Without cover-glass.—Sputum can be spread and stained on the glass slide without the use of a cover-glass, the oil of cedar being placed directly on the stained sputum, and the oil immersion lens dipping into it. It is a rapid and cheap way; and when a given case is to be studied daily the method is useful.

Pure Cultures from Sputum.—Kitasato recommends the thorough washing, changing the water ten times, of the small masses found in the sputum of tubercular persons. When such specimens are examined they show tubercle bacilli alone, and when inoculated in agar give rise to pure cultures.

Staining Bacillus Tuberculosis in Tissue (sections).—The general method of Gram can be used, but the better way is to use the following:—

Carbol-fuchsin, 15 to 30 minutes.
5 per cent. sulphuric acid, 1 minute.
Alcohol, until a light-red tinge appears.
Weak methylin blue, 3 to 5 minutes.
Alcohol, for a few seconds.
Oil of cloves, until cleared.
Canada balsam, to mount in.

Instead of carbol-fuchsin, *alcoholic solution of fuchsin* or *aniline water fuchsin* can be used, but the sections must remain in the stain over night.

Hardened sputum and sectioning.—Sputum can be hardened by placing it in 98 per cent. alcohol. Thin sections can be obtained by imbedding the hardened sputum in collodion. The sections are then stained as ordinary tissue sections.

To preserve sputum.—Sputum can be preserved for future use by placing it in alcohol, where it can be kept for months. Cover-glass preparations can then be made by softening the coagula with a small amount of liquor potassa.

The resisting action of the bacillus to acids is supposed to be due to a peculiar arrangement of the albumen and cellulose of the cell rather than to any particular capsule around it.

Pathogenesis.—When a guinea-pig has injected into its peritoneal cavity some of the diluted sputum containing tubercle bacilli it perishes in about three weeks, and the following picture presents itself at the autopsy: at the point of inoculation a local tuberculosis *shows itself*, little tubercular nodules containing the characteristic bacilli. In the lungs and the lymphatics, similar tubercles are found, a general tuberculosis.

If the animal lingers a few weeks longer, the tubercles become necrosed in the centre and degeneration occurs, the periphery still containing active bacilli, cavities having formed in the centre.

Since the bacilli die in course of time, killed by their own products, their number forms no correct guide of the damage present.

Even their absence in the sputum does not preclude the absence of a tubercular process. It is their presence only that warrants a positive declaration.

They are found in the blood only when a vessel has come in direct contact with a tubercular process through rupture or otherwise. They have been found in other secretions, milk, urine, etc.

Man is infected as follows :—

Through wounds.—Local tuberculosis.

Through nutrition.—Milk and meat of tuberculous animals.

Phthisical patients swallowing their own sputum and causing an intestinal tuberculosis.

Inhalation.—This is the most usual way, probably constituting the cause in $\frac{9}{10}$ of the cases.

The sputum of phthisical patients expectorated on the floors of dwelling-houses in handkerchiefs, etc., dries, and the bacilli set free are placed in motion by the wind or rising with the dust are thus inhaled by those present. When the sputum is kept from drying by expectoration in vessels containing water, this *great danger* can be *avoided.*

Nearly all the cases of *heredity* can be explained in this manner. The young children, possessing very little resistance, are constantly exposed to the infection through *inhalation* and also by *nutrition.*

Immunity.—No one can be said to be immune, though persons who have been greatly weakened would offer less resistance than healthy individuals.

Products of Tubercle Bacilli. The last years have developed some wonderful facts in relation to this important deadly bacillus. In 1889, Dr. Dixon, then Professor of Hygiene at the University of Pennsylvania, spoke of a method of curing tuberculosis in guinea-pigs and with products obtained from the bacillus; not much was thought of this statement at the time.

In August, 1890, *Koch*, before the Medical Congress claimed that he also had been able to cure tuberculosis in guinea-pigs, and would be able to give some interesting facts later on. In November he claimed that he had obtained reactions in man similar to those in the guinea-pig, and believed that a cure was at hand.

In the excitement which followed this statement, the greatest hopes were raised and the impossible expected. In January, 1891, Koch made public the manner of preparing the lymph or

"Tuberculin" or "Kochin," as it was variously called: old cultures of tubercle bacilli mixed with 60 per cent. glycerine and filtered through a Chamberlain-Pasteur filter, the filtrate thus obtained being a dark-brown liquid, sp. gr. somewhat higher than water, an odor like "beef extract," a sweetish taste, not soluble in alcohol; according to Jollas, containing 50 per cent. water, and showing a strong Biuret reaction; 1 milligramme of the lymph is supposed to contain but $\frac{1}{15000}$ milligramme of the active principle. Dixon's lymph is obtained in a very similar manner, and no doubt contains the same principle.

Dixon recommends instead of using the pure culture for obtaining the lymph, the tuberculosis lung of calf, a portion of which is treated with water and glycerine, and then filtered through Chamberlain-Pasteur filter without pressure.

Manner of Using Koch's Lymph.—One milligramme of the Koch's lymph is injected under the skin of one suffering with a tubercular process, and in a few hours to a few days, a rise of temperature, tightness about the chest, and exaggerated coughing spells take place, the symptoms varying in intensity; usually a secondary rise occurs on the following day. The dose has been gradually increased until the reactions subsided, and 600 milligrammes have then been borne without any reaction.

On Lupus the process could be watched and was very characteristic; a peculiar redness after the first injection, and after a few more injections scabs formed, and an apparent cure seemed to be obtained, but relapses were common and but very few authentic cures if any can now be had.

Tuberculin is a protein, and has no action on the bacilli, but seems to act on tubercular tissue, adding to the inflammation and exciting phagocytosis.

Koch believed that the tuberculosis tissue was rendered necrotic by this toxic principle, making the soil unfit for the bacilli which then perished or were expectorated.

Virchow dampened the excitement and ardor by showing a great diffusion of fresh miliary tubercles in the bodies of persons who had died and who had been treated with the lymph. Cool, careful, and untiring study and time taken together will, we trust, bring a happy solution and a genuine remedy.

Tuberculocidin.—This is an albuminoid obtained from the original tuberculin by precipitation with alcohol. Klebs uses it as a cure for tuberculosis. The results are as yet undecided.

Tuberculin as a diagnostic agent.—In cattle it has been used in doses of 30 to 40 grammes. When they are tubercular severe reactions follow its injection.

Lepra Bacillus. (Hansen.)

Origin.—In 1880 Armauer Hansen declared, as the result ot many years' investigation, that he found a bacillus in all leprous processes.

Form.—Small slender rods somewhat shorter than tubercle bacilli, otherwise very similar in appearance.

In the interior of the cell two to three oval spaces are usually seen, not known if spores or otherwise.

Properties.—They are immotile, do not liquefy the nutrient media.

Growth.—Bordoni-Uffreduzzi have obtained growths upon blood serum to which peptone and glycerine had been added. The growth is very slow, requiring about eight days at a temperature of 37° C.

Colonies.—Small grayish round spots, under microscope appearing like a close-netted spider web around a firm centre.

Stab Cultures.—Show a waxy-like growth along the needle track.

Staining.—They resist the decolorizing action of acids as the tubercle bacilli, but they are easily stained, requiring but a few minutes with the ordinary watery solutions. They take Gram's stain readily.

Pathogenesis.—Arning has inoculated prisoners with tissue obtained from leprous patients, and produced true leprosy.

Rabbits which had been infected through the anterior chamber of the eye showed the lepra nodules (containing the lepra bacilli) diffused through various organs.

In man the skin and peripheral nerves are principally affected, but the lymphatic glands, liver, and spleen can also become the seat of the lepra nodules. The lepra cells which compose these nodules contain the bacilli in large numbers. By applying a vesicant to the leprous skin the serum thereby obtained will contain great numbers of bacilli. This is a simple diagnostic test.

Method of Infection.—Not yet determined; the air, soil, water, and food of leprous districts have been carefully examined without result.

Syphilis Bacillus of *Lustgarten* (Smegma Bacillus of Alvarey and Tavel). Lustgarten in 1885, through a certain staining process, found peculiar bacilli in syphilitic tissues which he thought had a direct connection with the disease.

In the same year Alvarey and Tavel and Matterstock found a similar bacillus reacting in the same way to Lustgarten's color method in normal secretions, especially in the smegma of the prepuce.

The question yet remains an open one, what relation the syphilis or the smegma bacillus bears to syphilis, and will remain so until the bacillus can be cultivated, which so far has not been accomplished.

Origin.—In the cells of syphilitic tissue, in the secretion of syphilitic ulcers, and in the smegma of the prepuce and vulva.

Form.—Small slender rods similar in appearance to tubercle bacilli, sometimes swelled at the ends and curved S-shaped.

Colorless oval spaces also present, which Lustgarten calls spores.

Growth.—As before mentioned, they have not yet been cultivated.

Staining.—Lustgarten's method :—

1. Aniline water gentian violet, 12 to 24 hours, and then 2 hours longer in brood oven.
2. Rinsed in alcohol. 2 to 3 minutes.
3. Aqueous solution potass. permang. (1½ per cent.). 10 seconds.
4. Aqueous solution of sulphuric acid. 2 seconds.
5. Aq. destil. to wash.

Numbers three, four, and five repeated, until the section is colorless. Then alcohol, oil of cloves, and Canada balsam as usual.

De Giacomi's method :—

1. Aniline water fuchsin. 24 hours.
2. Rinse in dilute tr. ferri chlor. sol.
3. Decolorize in concentrated tr. ferri chlor.
4. Wash in alcohol, oil of cloves, Canada balsam, etc.

For cover-glass preparations, wash in water instead of alcohol.

Tubercle and lepra bacilli are colored by this method also, but syphilis bacilli become decolorized if washed with acids.

Pathogenesis.—No pathological actions have yet been definitely proven. They are found in greater quantities the younger the infection is.

Fig. 54.

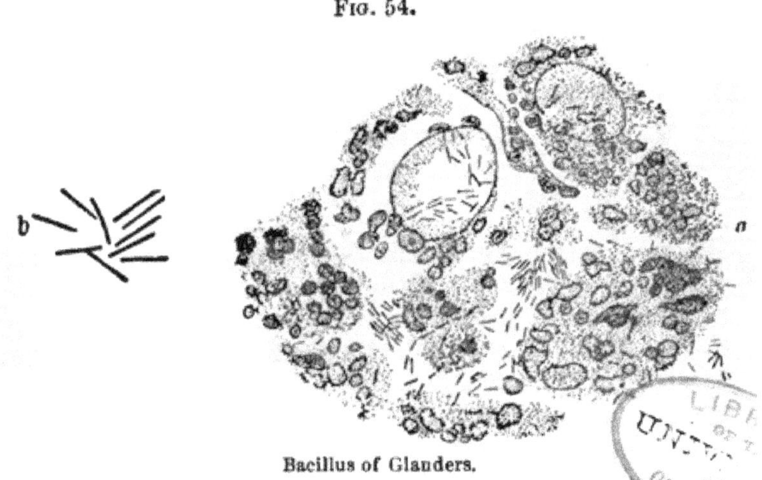

Bacillus of Glanders.

Bacillus of Glanders. (*Bacillus Mallei*, Löffler-Shütz.) *Rotz bacillus.*

Origin.—In the "farcy buds" or little nodules of the disease, by Löffler and Shütz in 1882.

Form.—Small slender rods, about the size of the tubercle bacillus. The ends rounded. Never appearing in large collections, usually singly. *Spores* are present.

Properties.—The rods are very resistant, living in a dried state for three months and longer without any spores present. They are not motile; possess, however, great molecular vibration.

Growth.—The growth occurs between 25° and 40° C., best at 37° C.; it is very sparse upon gelatine, but on glycerine-agar or blood serum a very abundant growth occurs.

Colonies.—On agar or glycerine-agar there appear in two to three days small white glistening drops, which under microscope seem as round granular masses with an even periphery.

Stroke Cultures.—On glycerine-agar and blood serum small transparent drops of whitish or grayish color, which soon coalesce to form a broad band.

Potato.—An amber-colored honey-like growth which gradually turns red.

Staining.—Since the bacillus is very easily decolorized, some special methods have been recommended.

Löffler's.—(For cover-glass preparations.)

1. Alkaline methylin blue (Löffler's). 5 minutes.
2. Acetic acid with a few drops of tropæolin. 1 second.
3. Washed in water.

For Sections.—Instead of tropæolin acetic acid, the following mixture is used:—

℞—Oxalic acid 5 per cent.	. . .	gtt. j.
Conc. sulphuric. acid.	. . . . .	gtt. ij.
Aq. destill.	. . . .	℥ij.—M.

The sections are kept in this 5 seconds.

Kühne's method. *Coverglass.*

1. Warm carbol-blue 2 min.
2. Decolorized in weak sol. of muriatic acid (10 parts to 500).
3. Washed in water.

Sections of Tissue.

1. Carbol-blue, ½ hour.
2. Decolorized in ½ per cent. muriatic acid.
3. Washed in distilled water.
4. Dehydrated in alcohol 1 second.
5. Aniline oil with 6 gtts. of turpentine. 5 min.
6. Turpentine, xylol, Canada balsam.

If contrast stain, add 5 gtts. of safranin (Bismark-brown) to turpentine, and use it after the xylol.

Pathogenesis.—If horses, field mice, or guinea-pigs be inoculated subcutaneously, with but a very small quantity of culture, a local affection results, followed some time after by a general disturbance; ulcers form at the point of inoculation; little nodules, which then caseate, leaving scars and involving the lymphatics; metastatic abscesses then occur in the spleen and lungs, and death arises *from exhaustion.* Cattle, pigs, and rabbits are not easily affected; man is readily attacked.

Manner of Infection.—Glanders being a highly contagious disease, it requires but a slight wound to allow it to gain entrance.

In horses the primary sore seems to be at the nasal mucous membrane. In man it is usually on the fingers. Boiling water or 1-10,000 sublimate solution will quickly destroy the virulence of this bacillus.

Bacillus of Diphtheria. (Klebs-Löffler.)

Origin.—In diphtheritic membrane, by Löffler, in 1884.

Form.—Small, slightly curved rods about as long as tubercle bacilli and twice as broad; the ends are at times swollen; spores have not been found.

Properties.—They do not possess any movement; do not liquefy gelatine. They are not very resistant, being destroyed by a temperature of 50° C., but they have lived on blood-serum five months.

Growth.—Grow readily on all media, between temperature of 20° and 40° C. They are facultative anærobic; they grow quite rapidly and profusely. Egg cultures (Hueppe's method) give good growths.

Colonies on Gelatine Plates.—At 24° C. little round colonies, under low-power, granular centre; irregular borders.

Stab Cultures.—Small, white drops along the needle track. In glycerine-agar a somewhat profuse growth.

Potato.—On alkaline surface, a grayish layer in 48 hours.

Blood-Serum (after Löffler).—Blood serum 3 parts, and bouillon 1 part; the bouillon contains peptone, 1 per cent., chloride of sodium, ½ per cent., and dextrin, 1 per cent.

On this medium a very thick yellowish-white layer occurs on the surface, and isolated colonies in the upper strata.

Staining.—Is not colored by Gram's method. Stained best with Löffler's alkaline methylin-blue.

Pathogenesis.—By inoculation, animals, which naturally are not subject to diphtheria, have had diphtheritic processes develop at the site of infection; hemorrhagic œdema then follows, and death.

In rabbits paralyses develop, and when the inoculation occurs upon the trachea, all the prominent symptoms of diphtheria show themselves.

Manner of Infection in Man.—The exact way is not yet known. It is supposed that the mucous membrane altered in some manner, the diphtheria bacillus, then gains entrance and the disease develops.

Products.—But it is not the mere presence of the bacillus that gives rise to all trouble; certain products which they generate get into the system and produce the severe constitutional symptoms.

Roux and Yersin, in 1888, discovered that the injection of the filtered culture bouillon (that is, freed of all diphtheria bacilli) gave rise to the same palsies as when the bacilli themselves were introduced.

Toxalbumen of diphtheria.—Brieger and Fränkel filter the bouillon culture, evaporate (in vacuo at 27° C.) to ⅓ volume, then treat with 10 volumes of alcohol and acetic acid the precipitate redissolved in water and reprecipitated with the acidulated alcohol until a clear aqueous solution is obtained; this is then dialyzed for 72 hours, and again precipitated with alcohol, and dried; a white amorphous body results, giving all the reactions of an albumen, and called by them toxalbumen.

Immunity.—Brieger and Fränkel, by injecting 10 to 20 c.cm. of a three weeks' old culture of diphtheria bacilli, which had been heated at 70° C. for one hour, produced an immunity in guinea-pigs against the virulent form.

Behring found several ways to make animals immune. One method was to infect them with diphtheria and then inject trichloriodine into them, which prevented them from dying, and they were then immune.

Site of Bacilli.—Bacilli are usually found in the older portions of the pseudo-membrane very near to the surface. The secretions of the throat of a diphtheritic child produced bacilli three weeks after the temperature was down to normal.

Streptococcus in Diphtheria. Streptococci have been found quite constant in diphtheria, but they resemble the streptococcus pyogenes, and have no specific action.

Bacillus of Typhoid or Enteric Fever. (Eberth-Gaffky.)

Origin.—Eberth found this bacillus in the spleen and lymphatic glands in the year 1880, and Gaffky isolated and cultivated the same four years later.

Form.—Rods with rounded ends about three times as long as they are broad. Usually solitary in tissue-sections, but in artificial cultures found in long threads. Flagella on the side.

Properties.—They are very motile; they take the aniline dyes less deeply than some similar bacilli. *Spores* have not yet been found; small oval spaces appear in some of the degenerated bacilli just at one end, but these bacilli are less resistant than those without this so-called spore; they do not liquefy gelatine.

Fig. 55.

Typhoid fever bacillus in pure culture. 650 diameters.

Fig. 56.

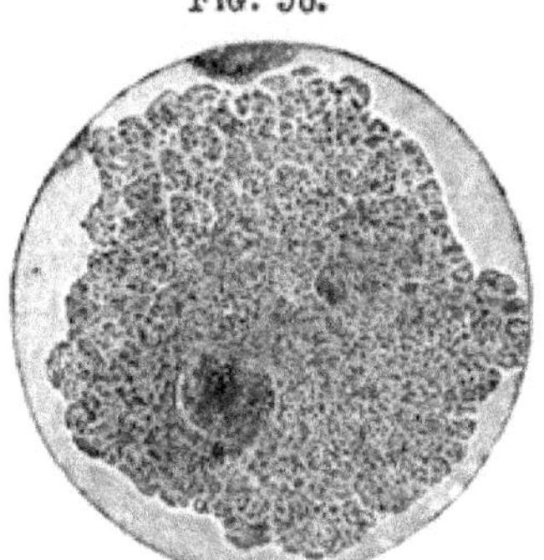

Colonies of typhoid bacilli 3 days old 100 ×. (Fränkel and Pfeiffer.)

Growth.—They are facultative anærobic; grow best at 37° C., but can also develop at ordinary room temperature. All nutrient media can be used as culture ground. They develop chiefly on the surface, and very slowly.

Colonies on Gelatine Plates.—Two forms; the ones near the surface spread out like a leaf, transparent with bluish fluorescence. The deeper ones resemble whetstone crystals of uric acid, the same yellowish tinge.

In five days they attain to 3 millimetres in diameter.

On Potato-Gelatine.—The colonies do not have the yellow color, they are transparent, later on they become dark brown with green iridescence.

Stab Cultures.—Mainly on the surface a pearly layer.

Stroke Cultures.—A transparent thick layer.

Potato.—The growth here is quite characteristic. At 37° C. in 48 hours a moist transparent film is formed over the whole surface, but so transparent that it can hardly be seen without close observation. If a small portion of this is placed under a microscope, it will be seen swarming with bacilli.

The growth never becomes more prominent; the potato must have a neutral or acid reaction.

Milk.—They grow very well in milk without producing any visible changes in its composition.

Carbolized-Gelatine.—Gelatine which has added to it $\frac{1}{10}$ per cent. carbolic acid will allow the typhoid bacillus to develop, other similar bacilli being destroyed.

Staining.—Colored with the ordinary aniline dyes, when they are *warmed;* since they are easily decolorized, acids should be avoided.

Gram's method is not applicable. Tissue sections stained as follows:—

Alkaline blue	1 hour.
Alcohol	5 seconds.
Aniline oil	5 minutes.
Turpentine oil	1 minute.
Xylol and Canada bals.	

Such a specimen should first be examined with low power, to focus little colored masses, then examined with immersion lens; these masses will be found composed of bacilli.

Similar Bacteria. The *Neapolitanus* bacillus of Emmerich or *fæces bacillus* of Briëger resembles the typhoid bacillus in many ways, the colonies being the same and its structure similar. But the growth on potato is very different; a thick, yellow, pasty layer is formed thereon.

In Water. Bacilli have been found which also resemble typhoid bacilli, and one must be very careful not to make any positive statement.

Examination of Water for Typhoid Bacilli.—When a water is supposed to contain typhoid bacilli, 500 c.cm. of the same is mixed with 20 gtts. of ¼-per cent. carbolic acid, which destroys many of the saprophytes.

Plates are then made as described under Water Analysis.

Those colonies which then form and have a tendency to liquefy, are touched on second day with permanganate of potassium, and when so colored, destroyed with bichloride of mercury.

Those that now develop are transferred by inoculation to fresh plates. At the end of eight days they are examined under microscope; every colony not possessing motile bacilli is discarded. The motile bacilli are tested with Gram's method of staining; those that do not take the stain are alone retained. Cultures are made from these upon potatoes, and, if the characteristic growth occurs, then only can they be called typhoid bacilli with any certainty.

Pathogenesis.—Lower animals have not yet been given enteric fever, though their death has been caused by injection of the bacilli into the veins of the ear.

In man it has been found in the urine, blood, sputum, milk, intestinal discharges, roseolar spots, and in various organs, as spleen, liver, lymphatic glands, and intestinal villi.

It is found in secretions several days after the attack has subsided. It is found only in this disease, and regularly.

Way of Infection.—The bacilli in the dejecta of the diseased person find their way into drinking water, milk, or dirty clothes, and so into the alimentary tract of a person predisposed to the disease. They enter the blood through the lymphatics, and so become lodged in various organs.

Products.—Briëger found a ptomaïne in the cultures which he named typhotoxin with the formula $C_9H_{17}NO_2$. It has no specific action. A toxalbumen insoluble in water has also been isolated, but, as experiment animals are immune to the disease, no definite actions have yet been determined.

The cultures, when old, show an acid reaction.

Bacillus Neapolitanus. (Emmerich.)

Origin.—During the cholera epidemic in Naples, in 1884, Emmerich found this bacillus in the blood and intestinal discharges of cholera-suffering patients. He supposed it to be the real cause of cholera; but since then it has been shown to be nothing more than the *Fæces bacillus* which Briëger described, and which is found in fæces of healthy persons, in the air and various putrefactive processes.

Form.—Very much like the typhoid bacillus, short rods with rounded ends with oval spaces in them as the typhoid.

Properties.—*Immobile*, differing thus markedly from typhoid. *Do not liquefy* gelatine.

Growth.—They are facultative anærobic; they grow more rapidly than the typhoid, and endure cold and heat better than they do.

Colonies.—They are exactly the same as typhoid—the same whetstone-shaped deep ones and the leaf-shaped surface ones.

Potato.—A thick yellow-brown pasty layer is formed instead of the transparent almost invisible growth of the typhoid bacillus.

Staining.—Do not take Gram. Fuchsin stains them easily.

Pathogenesis.—When large quantities injected into guinea-pigs, they die at times, sometimes with intestinal symptoms, sometimes without.

Bacillus Coli Communis. (Escherich.)

Found in human feces, intestinal canal of most animals, in pus and water.

Form.—Short rods with very slow movement, often associated in little masses resembling the typhoid germ.

Properties.—Does not liquefy gelatine, causes fermentation in saccharine solutions in the absence of oxygen, produces acid fermentation in milk.

Growth.—On potato a thick, moist, yellow-colored growth. Very soon after inoculation on gelatine a growth similar to typhoid. It can also develop in carbolized gelatine, and withstands a temperature of 45° C. without its growth being destroyed.

Pathogenesis.—Inoculated into rabbits or guinea-pigs, death follows in from one to three days, the symptoms being those of diarrhœa and coma; after death tumefactions of Peyer's patches and other parts of the intestine; perforations into peritoneal cavity, the blood containing a large number of germs.

Staining.—Ordinary stains; do not take Gram.

Site.—The bacillus has been found very constant in acute peritonitis and in cholera nostra. Its presence in water would indicate fecal contamination.

The growth on potato, the effect on animals, and its action towards milk are points of difference from the typhoid bacillus.

CHAPTER III.

PATHOGENIC BACTERIA—CONTINUED.

Spirillum Choleræ. (Koch.) *Comma bacillus of cholera.*

Origin.—Koch, as a member of the German expedition sent to India, in 1883, to study cholera, found this micro-organism in the intestinal contents of cholera patients, and by further experiments identified it with the disease.

Fig. 57.

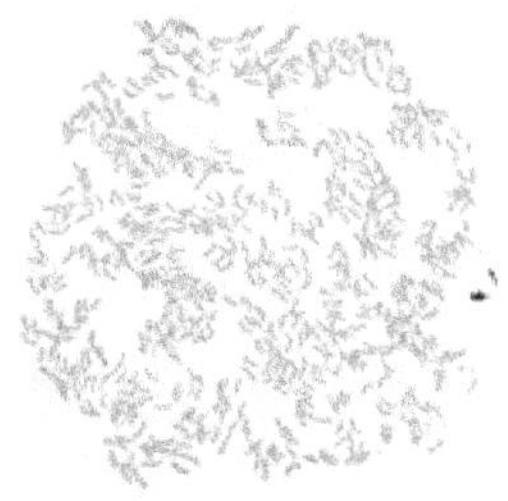

Comma bacillus, pure culture. 600 diameters.

Form.—The microbe as seen ordinarily appears as a short, arc-like body, about half the size of a tubercle bacillus, but when seen in large groups, spirals are formed, each little arc appearing then as but a segment, *a vibrio; each arc* is about three times as long as it is broad, and possesses at each end a *flagella.*

Properties.—They are very motile; liquefy gelatine. They are easily affected by heat and dryness. Spores have not been found, though some (Hüppe) claim arthrospores.

Growth.—Develops at ordinary temperatures on all nutrient media that have an alkaline or neutral reaction. They are facultative anærobic.

Colonies, gelatine.—After 24 hours, small white points which gradually come to the surface, the gelatine being slowly liquefied, a funnel-shaped cavity formed holding the colony in its narrow part, at the bottom, and on the fifth day all the gelatine is liquid. If the colonies of three days' growth are placed under microscope they appear as if composed of small bits of frosted glass with sharp irregular points.

Stab Culture.—After 30 hours a growth can be distinguished along the needle track, and on the surface a little cavity has been formed, filled up by a bubble of air, and this liquefaction proceeds until on the sixth day it has reached the sides of the tube, tapering, funnel-shaped to the bottom of the tube. After several weeks the spirilla are found in little collections at the bottom of the fluid gelatine. In eight weeks the bacilli have perished.

FIG. 58.

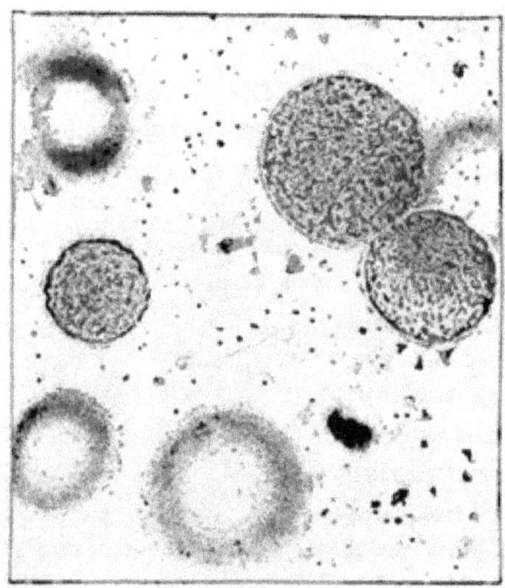

Cholera colonies after 30 hours 100 ×. (Fränkel and Pfeiffer.)

Agar.—Stroke cultures. A shiny white layer lasts many months.

Potato.—A yellow honey-like transparent layer, if the potato is kept at animal heat.

Bouillon. — A wrinkled scum is soon formed in bouillon. They live well and grow in sterilized milk and sterilized water, remaining virulent in the latter for many months. In ordinary water, the bacteria present are destructive to the comma bacillus, and they die in a few days.

Staining.—They are colored well with watery aniline solutions. The flagella can be well seen by staining according to the flagella stain.

Pathogenesis.—Experiment animals are not subject to cholera Asiatica, but by overcoming two obstacles Koch has produced choleraic symptoms in guinea-pigs. Nicati and Rietsch prevented peristalsis and avoided the acidity of the stomach juices by direct injection into the duodenum, after tying the gall-duct. Koch alkalinizes the gastric juice with 5 c.cm. of 5 per cent. sol. of sodii carbonas, and then injecting 2 grams of opium tincture for every 300 grams of weight into the peritoneal cavity paralyzes peristalsis. The cholera culture then introduced through a stomach-tube, the animals die in forty-eight hours, presenting the same symptoms in the appearance of the intestines as in cholera patients, the serous effusion containing great numbers of spirilla.

Manner of Infection in Man.—Usually through the alimentary tract, with the food or drink, the intestinal discharges of cholera patients having found entrance into the source of drinking water. Soiled clothes to fingers, fingers to the mouth, etc.; torpid catarrhal affection of the digestive tract predisposing. The microbe is not found in the blood or any organ other than the intestines, the tissue of the small intestines. It is also found in the vomit and the intestinal contents.

Fig. 59.

Comma bacillus in mucus, from a case of Asiatic cholera.

Products.—"*Cholera red.*" When chemically pure nitric or sulphuric acid is added to nutrient peptone cultures of the cholera bacillus a rose-red color is produced. This will not take place with other bacilli unless *nitrous acid is present.* The cholera bacillus forms nitrites from the nitrates present in the media, and also indol. The mineral acid splits the nitrites, setting free nitrous acid, which, with the indol, forms the red reaction. This pigment has been isolated and extracted and called "*cholera red.*" A ptomaïne, identical with cadaverin, and several other alkaloids have been obtained from the cultures. A toxalbumen and a toxicpeptone have lately been isolated, but no special actions ascribed to them.

Bacteria Similar to the Spirillum of Cholera.

Finkler-Prior Vibrio, or Spirillum Finkleri.

Origin.—Found in the intestinal contents of a patient suffering from cholera Asiatica in 1884, by Finkler and Prior, who thought it identical with the spirillum of cholera; it differs from it, however, in many ways, and has been found in healthy persons.

Fig. 60.

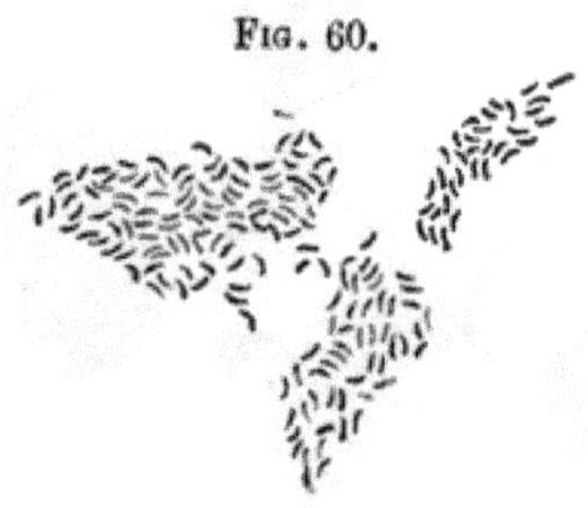

Spirillum Finkleri. 700 diameters. (Flügge.)

Form.—Somewhat thicker than the cholera vibrio, otherwise about the same form; it forms the long spirilla less often. Has *Flagella.*

Properties.—It is very motile. Liquefies gelatine in a short time.

Growth.—It grows quickly at ordinary room temperature. It is facultative ærobic.

Colonies on Gelatine Plates.—Round, finely granular colonies, which in twenty-four hours are ten times as large as the cholera colonies, and in forty-eight hours the whole plate is liquefied, it being then impossible to distinguish any separate colonies. The microscopic appearances in no way resemble the cholera colony.

Stab Cultures.—The gelatine is liquefied from above downwards, like a stocking in appearance, and in three days is completely liquid.

PLATE I.

TUBE CULTURES.

(From U. S. Government Report on Cholera.—*Shakespeare.*)

A.

Cholera Bacillus.
48 hours.
5% Gelatin.

B.

Cholera Bacillus.
60 hours.
5% Gelatin.

C.

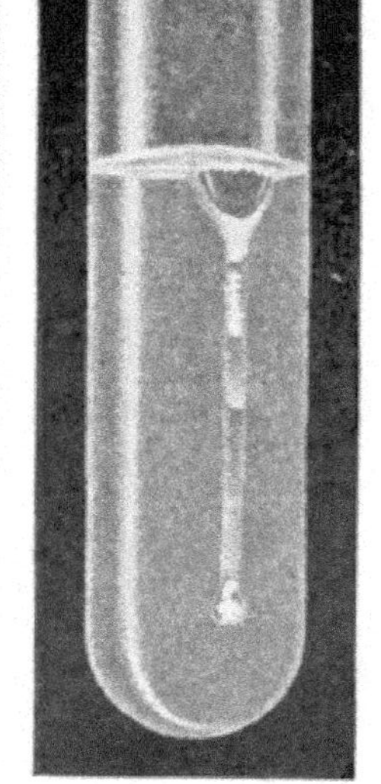

Cholera Bacillus.
72 hours.
15% Gelatin.

D.

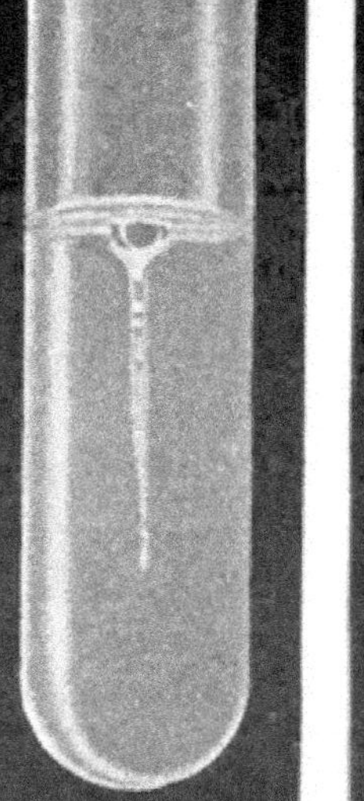

Deneke. Cheese Bacillus.
48 hours.
5% Gelatin.

E.

Deneke. Cheese Bacillus.
60 hours.
5% Gelatin.

Potato.—At ordinary temperature a thick gray layer covering the whole surface.

Water.—It soon perishes in water.

Staining.—Ordinary aniline dyes.

Pathogenesis.—For man it has no specific action. If it is injected into Guinea pigs, prepared as described under the cholera bacillus, they die, the intestines having a foul odor, and the bacilli then found in great numbers.

FIG. 61.

Stab Culture. (Finkler-Prior.)

Spirillum Tyrogenum. (Deneke.)

Origin.—In 1885 Deneke found in old cheese a spirillum very similar in appearance to the cholera spirillum.

Form.—The same as the cholera vibrio.

Properties.—Very motile, liquefy gelatine.

Growth.—They grow quicker than the cholera, and slower than the Finkler; they are also facultative aërobic.

Colonies.—They at first resemble cholera colonies; they have, however, a yellow-green iridescence, and are somewhat more irregular; also grow more rapidly.

Stab Cultures.—A thick line along the needle-track and the yellow colonies forming at the bottom, on the surface a bubble of air similar to the cholera. The gelatine is all liquid in two weeks.

Potato.—At brood-heat a thin yellow membrane, but not always constant.

Staining, as cholera bacillus.

Pathogenesis.—When injected into animals prepared as for the cholera bacillus, a certain number die.

Vibrio Metschnikovi. (Gamaleia.)

Origin.—In the intestines of fowls suffering from a gastro-enteritis, common in Russia. Gamaleia found a spirillum which bears so close a resemblance to the cholera bacillus, both in form and growth, that it cannot be distinguished by these characteristics alone.

Form.—As cholera bacillus.

Growth.—Two kinds are found on the gelatine plate—one that is identical in appearance with the cholera colony, the other more liquefying, resembling the Finkler spirillum. If now a second plate be inoculated from either one of these forms, both kinds again are found grown, so that it is not a mixture of two bacilli.

Stab Culture.—Similar to the cholera growth, a trifle faster in growing.

Staining.—As cholera.

Pathogenesis.—To differentiate it from cholera, these bacilli, when injected into animals, prove very fatal, and no especial precautions need be taken to make the animal susceptible. In the pigeon, guinea-pig, and chicken it produces a hemorrhagic œdema, and a septicæmia which has been called "*Vibrion septicæmia.*" The blood and organs contain the spirilla in great numbers.

Products.—The nitrites are formed just as in cholera bacillus, and the red reaction given when mineral acids added to gelatine cultures. Certain products also which, when injected, give immunity. The cultures are first heated for one half hour at 100° C., which destroys the germs, and then this sterilized product injected. (5 c.cm. of a five days' old sterilized culture.)

In a couple of weeks 1 to 2 c.cm. of the infected blood can be injected without causing any fatal result.

Bacteria of Pneumonia. Two forms of bacteria have been found in this disease, and thought at different times to be the cause of the same.

Neither one of them is constant in pneumonia; and since many other pathological processes have shown them they can hardly be set down as the sole cause of pneumonia.

Klebs in 1875 called attention to the presence of bacteria in pneumonia, and in 1882 Friedlander developed a bacillus from the lung tissue of a pneumonic person, which he thought was a coccus, and called it pneumococcus.

In 1886 A. Fränkel and Weichselbaum proved that this microbe was not constant, in fact was rare.

A. Fränkel obtained in the majority of cases of pneumonia a microbe that he had described in 1884 under the name of sputum-septicæmia micrococcus.

Weichselbaum now called it "*Diplococcus Pneumonia*," and believed it to be the real cause of pneumonia. It has been found in many other serous inflammations, and also in the mouth of healthy persons.

Streptococcus pyogenes and *staphylococcus pyogenes aureus* have been found in some cases.

Fig. 62.

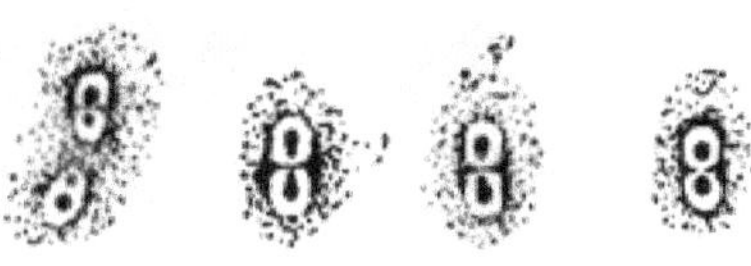

Pneumo-bacillus of Friedlander, with capsule.

Pneumo-bacillus (Pneumococcus). (Friedlander.)

Origin.—In the lung of a croupous-pneumonia person, by Friedlander, in 1882.

Form.—Small, almost oval-shaped rods, nearly as wide as they are long; often in pairs, they were at first believed to be cocci. In bouillon cultures the rod-form becomes more visible. In tissues each bacillus is surrounded by a faint capsule; but not around those developed in artificial cultures. Spores have not been found.

Properties.—They are immobile; do not liquefy gelatine. A gas is produced in gelatine cultures.

Growth.—Grows rapidly on all media at ordinary temperature: is facultative ærobic.

Colonies.—On gelatine plates. Small white round colonies. reaching the surface in the course of three or four days; appearing then as little buttons, with a porcelain-like shimmer, the edges smooth.

Stab Culture.—A growth along the needle-track, but on the surface a button-like projection, which gives to the growth the

Fig. 63.

Bacillus of Pneumonia. Stab Culture. (*Nail Culture.*)

appearance of a *nail driven into the gelatine*, its head resting on the surface; therefore such cultures are called "*Nail cultures.*" See Fig. 63. Old cultures are colored brown, and contain bubbles of gas.

Potato.—A yellow, moist layer in a few days at brood-heat. Gas bubbles develop.

Staining.—The ordinary aniline stains. The sections do not take Gram's method; are therefore not suited for double staining.

Capsule.—Stained as follows:—

Cover glasses.

1. Acetic acid, two minutes.
2. Allow acetic acid to dry by blowing air upon it through a glass tube.
3. Saturated, aniline water. Gent. violet, ten seconds.
4. Rinse in water. Mount in Canada balsam.

For Sections.

℞

1. Stain in warm { conc. alc. gent. violet, 50.0; aqua, 100.0; acetic acid, 10. } for 24 hours. M.

2. Rinse in one per cent. acetic acid.
3. Alcohol to dehydrate. Mount in balsam.

The capsule will be found stained a light blue, the bacillus a deep blue.

Pathogenesis.—Animals are not affected unless the culture is injected intrapleura.

Pneumobacillus of Frankel. (A. Fränkel and Weichselbaum.)

Synonyms.—Pneumococcus; Diplococcus of Pneumonia; Micrococcus of sputum septicæmia; Micrococcus Pasteuri; Diplococcus lanceolatus.

Origin.—A. Fränkel found it in the sputum of pneumonic patients, thinking it at first to be the micrococcus of sputum septicæmia; later he believed it to be the cause of pneumonia.

Form.—Oval cocci they were at first called, but they are now known to be rod-shaped, being somewhat longer than broad; varying, however, much in size and shape. Usually found in pairs, sometimes in filaments of three and four elements. In the material from the body a capsule surrounds each rod. In the artificial cultures this is not found.

FIG. 64.

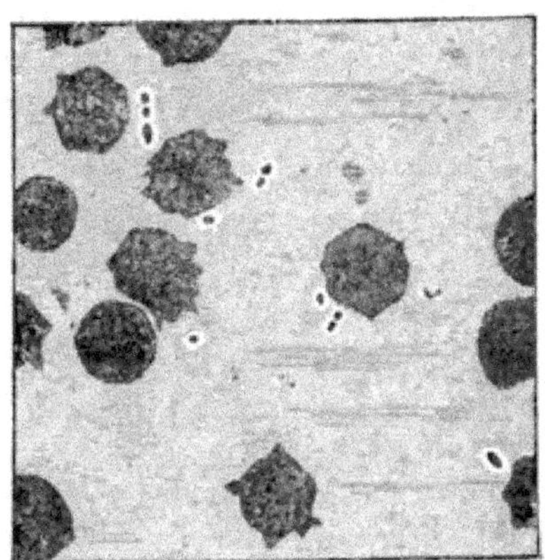

Bacillus of pneumonia in blood of rabbit 1000 ×. (Fränkel and Pfeiffer.)

Properties.—They are without self-movement; do not liquefy gelatine.

Growth.—Grow only at high temperature, 35° C.; are facultative anærobic. The culture media must be slightly alkaline; the growth is slow.

Colonies on Gelatine Plates.—Since the temperature must be somewhat elevated, the gelatine media need to be thicker than usual (15 per cent. gelatine), in order to keep it solid, and a temperature of 24° C. used. Little round white colonies, somewhat granular in the centre, growing very slowly.

Stab Cultures.—Along the needle-track small separate white granules, one above the other, like a string of beads.

Stroke Culture.—On agar, transparent, almost invisible little drops resembling dew moisture.

Bouillon.—They grow better here than in the other media, remaining alive a longer period of time.

Staining.—Takes Gram's method and the other aniline stains very readily. The *capsule* stained the same way as that of the *Friedlander bacillus.*

Pathogenesis.—Rabbits and guinea-pigs, if subcutaneously injected, die in the course of a couple of days with septicæmia. (0.1 c.cm. of a fresh bouillon culture suffices.)

Autopsy shows greatly enlarged spleen and myriads of bacilli in the blood and viscera, the lungs not especially affected. If injected per trachea, a pneumonia occurs. In man in 90 per cent. of croupous pneumonia they are found and usually only during the existence of the "prune juice" sputum, *i. e.*, the first stage.

Fig. 65.

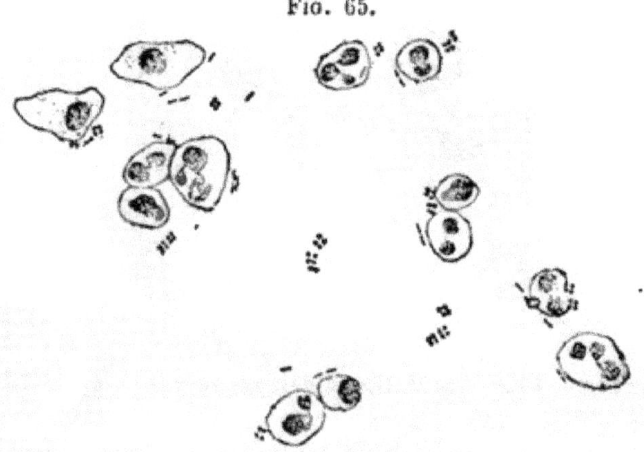

Micrococcus tetragenus in sputum (tubercle bacillus also).

They have also been found in pleuritis, peritonitis, pericarditis, meningitis, and endocarditis. They stand in some intimate relation with all infectious inflammations of the body. Their presence in healthy mouth secretion does not speak against it, it requiring some slight injury to allow this ever-present germ to develop the disease.

Anti-toxin of Pneumonia. (Klemperer.)

The injection of very diluted cultures of the virulent bacilli intravenously has produced an immunity in rabbits and guinea-pigs. The serum of such artificially immune animals when filtered through a Chamberland filter and injected into a rabbit suffering with pneumonia, cured the same; or when injected into a susceptible animal produced in it immunity very quickly. This principle is ascribed to an anti-toxin formed in the tissues by the diluted proteids, and this anti-toxin neutralizes the toxicity of the strong virus.

Bacillus of Rhinoscleroma. (Frisch. 1882.) It was found in the tissue of a rhinoscleroma, but resembles the Friedlander bacillus in nearly every respect, and as the disease rhinoscleroma was not reproduced by the inoculation of the bacillus in animals, it can be considered identical. The growth, cultures, and properties are the same as the pneumobacillus of Friedlander.

Micrococcus Tetragenus. (Koch. Gaffky).

Origin.—Koch found this microbe in the cavity of a tuberculous lung. Gaffky, in 1883, studied its pathogenic actions and gave it the name it now bears.

Fig. 66.

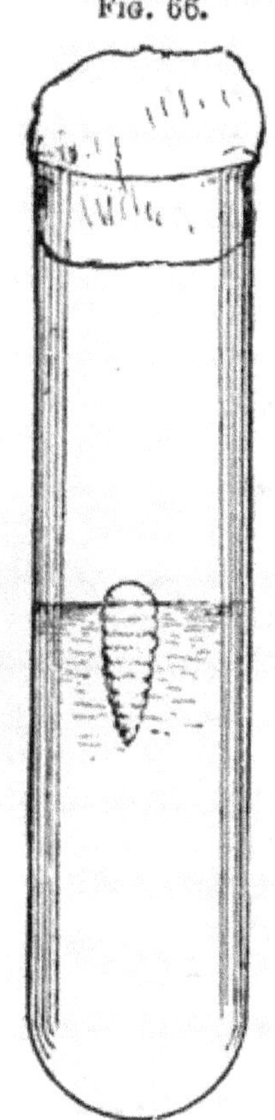

Stab Culture. Micrococcus tetragenus.

Form.—Cocci which are gathered in the tissues in groups of four, forming a square, a tetrad. See Fig. 65. In artificial culture, sometimes found in pairs. A capsule of light gelatinous consistence surrounds each tetrad.

Properties.—They are immobile; do not liquefy gelatine.

Growth.—They grow well on all nutrient media at ordinary and brood temperatures; are facultative ærobic. They grow slowly.

Colonies in gelatine plates. In two days, little white spots, which when on the surface form little elevations of a porcelain-like appearance; under low power they are seen very finely granulated.

Stab Culture.—Small round separated colonies along the needle-track, and on the surface a button-like elevation, a form of "nail culture." See Fig. 66.

Potato.—A thick slimy layer which can be loosened in long shreds.

Staining.—Colored with the ordinary aniline stains. Gram's method also applicable.

Pathogenesis.—White mice and guinea-pigs die in a few days of septicæmia when injected with the tetragenus cultures, and the micrococcus is then found in large numbers in the blood and viscera.

Field mice are immune.

In the cavities of tubercular lungs, in the sputum of phthisical and healthy patients, it is often found, but what action it has upon man has not yet been determined.

Capsule Bacillus. (Pfeiffer.)

Origin.—Stringy exudate and blood of a dead guinea-pig.

Form.—Thick little rods, sometimes in long threads. Large oval capsules in the stained preparations.

Properties.—Immotile, not liquefying, an odorless gas in gelatine cultures.

Growth.—At ordinary temperatures, quite rapidly; facultative anærobin.

Gelatine Plates.—Oval points, and like a porcelain button on the surface.

Stab Cultures.—Like the pneumonia bacillus of Friedlander.

Potatoes.—Abundant growth, yellow color and moist, coming off in strings.

Staining.—Hot fuchsin colors the capsule intensely; then carefully decolorizing with acetic acid, the capsules are seen red or light violet around the deeply-tinged bacillus. Gram's method not applicable.

Pathogenesis.—Subcutaneously injected in mice, they die in 48 hours. Rabbits die when a large quantity is injected into the circulation. The blood and juices have a peculiar stringy, fibrinous consistence.

Bacillus of Influenza. (Pfeiffer.)

A small bacillus about one-half the size of the bacillus of mouse septicæmia, and arranged in chain-form, is believed to be the cause of influenza. It develops upon blood-serum agar. It is aërobic. Without movement; does not take the gram stain. It is best stained with diluted carbol-fuchsin, the contrast-stain being Löffler's methylene-blue. It is found in the sputum and in the bronchial and nasal secretions of influenza patients.

Micro-Organisms of Suppuration. The suppuration of wounds is due to the presence of germs. The knowledge of this fact is the basis of the antiseptic treatment in surgery; for when the microbes can be destroyed or their entrance prevented, the wounds are made clean and kept without suppurating. Various forms of bacteria have been found in septic processes, and the formation of *pus* cannot be ascribed to any particular one alone ; some, more common than others, are found in nearly all forms of suppuration ; others give rise to special types.

Wounds are often irritated by foreign bodies and chemicals, and a discharge occurs in them even when every aseptic and antiseptic precaution has been taken ; but such a discharge is free from bacteria, and no more like pus than a benign growth is like a malignant one.

Streptococcus Pyogenes. (Rosenbach.) *Streptococcus erysipelatis.* (Fehleisen.)

Origin.—Fehleisen discovered this microbe in the lymphatics of the skin in erysipelas, and he thought it the cause of the same. Under the name streptococcus pyogenes, Rosenbach

Fig. 67.

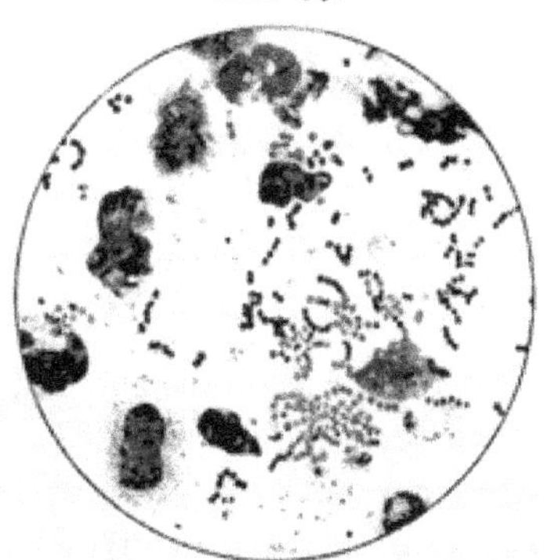

Streptococcus pyogenes in pus 1000 X. (Fränkel and Pfeiffer.)

described an identical coccus which has been found in nearly all suppurative conditions.

Form.—Small cocci singly and in chain-like groups. Spores have not been found, though it is supposed because of their permanency that spores are present.

Properties.—They are immotile, do not liquefy gelatine.

Growth.—They grow slowly, usually on the surface, and best at higher temperatures.

Colonies.—In three days a very small grayish speck, which hardly ever becomes much larger than a pin-head ; under microscope, looking yellowish, finely granular, the edges quite defined.

Stab Cultures.—Along the needle-track little separated colonies like strings of beads, which after a time become one solid white string.

Stroke Culture.—Little drops, never coalescing, having a bluish tint.

Potato.—No apparent growth.

Bouillon.—At 37° C. clouds are formed in the bouillon, which then sink to the bottom, and long chains of cocci found in this growth.

Staining.—Easily colored with the ordinary stains. Gram's method is also applicable.

Pathogenesis.—Inoculated subcutaneously in the ear of a rabbit, an erysipelatous condition develops in a few days, rapidly spreading from point of infection.

In man, inoculations have been made to produce an effect upon carcinomatous growths. *Erysipelas* was always produced thereby. When it occurs upon the valves of the heart, *endocarditis* results. *Puerperal fever* is caused by the microbe infecting the endometrium, the *Streptococcus puerperalis* of Fränkel being the same germ.

In scarlatina, variola, yellow fever, cerebro-spinal meningitis, and many similar diseases, the microbe has been an almost constant attendant.

In erysipelas the cocci reside in the lymphatic glands and ducts. They have not been found in the blood. In air, soil, and putrefying matters they have been often discovered.

Staphylococcus Pyogenes Aureus. (Rosenbach.)

Origin.—Found very commonly in pus (80 per cent. of all suppurations), in air, water, and earth; also in sputum of healthy persons.

FIG. 68.

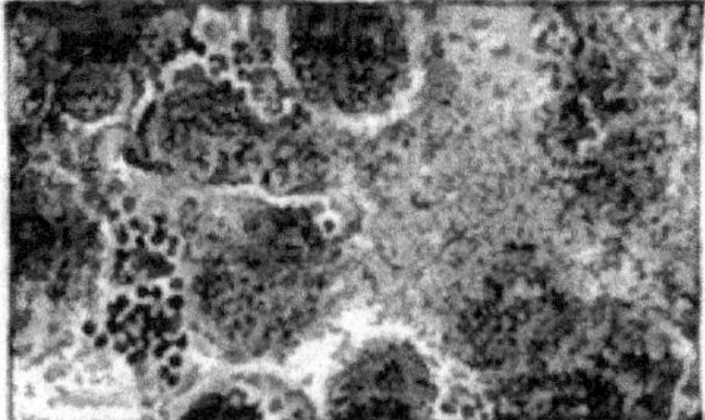

Staphylococcus pyogenes aureus in pus 1000 X. (Fränkel and Pfeiffer.)

Form.—Micrococci in clusters like bunched grapes, hence the name staphylo, which means grape. They never form chains. Spores have not been found, though the cocci are very resistant.

Properties.—Without movement; liquefying gelatine. It gives rise to an orange-yellow pigment in the various cultures.

Growth.—It grows moderately fast at ordinary temperature, and can live without air, a facultative ærobin and anærobin.

Colonies on Gelatine.—On second day small dots on the surface, containing in their centre an orange-yellow spot. The gelatine all around the colony is liquefied; the size is never much greater than that attained the second day.

Colonies on Agar.—The pigment remains a long time.

Stab Culture.—At first, gray growth along the track, which, after three days, has settled at the bottom of the tube in a yellow granular mass, the gelatine being all liquid.

Stroke Culture on Agar.—The pigment diffused over the surface where the growth is, in moist masses.

Potato.—A thin white layer which gradually becomes yellow and gives out a doughy smell.

Staining.—Very readily colored with ordinary stains; also with Gram's method.

Pathogenesis.—When rabbits are injected with cultures of this microbe into the knee-joint or pleura, they die in a day. If injected subcutaneously, only a local action occurs, namely, abscesses.

FIG. 69.

Stab culture. Micrococcus pyogenes aureus.

If directly into circulation, a general phlegmonous condition arises, the capillaries become plugged with masses of cocci, infarct occur in kidney and liver, and metastatic abscesses form in viscera and joints. Garré, by rubbing the culture on his forearm, caused carbuncles to appear.

Fracturing a long bone in an animal and then injecting the staphylococcus into a large vein, as the jugular, will produce osteomyelitis. Becker isolated this microbe from several cases of osteomyelitis, and thought it a specific germ, giving it the name of "*micrococcus of osteomyelitis.*"

Suppuration is nearly always produced by this microbe, and it is found in the majority of suppurative processes.

Micrococcus Pyogenes Albus. (Rosenbach.) Similar in every respect to the pyogenes-aureus, except that it does not form a pigment.

Micrococcus Pyogenes Citreus. (Passet.) This staphylococcus liquefies gelatine less rapidly than the pyogenes aureus, and forms a citron-yellow pigment instead of the orange-yellow of the aureus.

Micrococcus Cereus Albus. (Passet.) Differs from the pyogenes albus in the form of colony. A white shiny growth like drops of *wax;* hence the name *cereus.* It was found in pus, but gave no action in animals.

Micrococcus Cereus Flavus. (Passet.) A lemon-yellow colored growth after a short time, otherwise not differing from *cereus albus.*

Micrococcus Pyogenes Tenuis. (Rosenbach.)

Origin.—Found in the pus of large inclosed abscesses.

Form.—Cocci, without any especial arrangement.

Properties.—Not much studied.

Growth.—It was cultivated on agar, on which it formed in clear thin colonies; along the needle-track an opaque streak, looking as if varnished over.

Bacillus Pyocyaneus. (Gessard.)

Synonyms.—Bacterium æruginosum, bacillus fluorescens. (Schröter.) The bacillus of bluish-green pus.

Origin.—Found in 1882 in the green pus in pyocyæmia.

Form.—Small slender rods with rounded ends, easily mistaken for cocci. Often in groups of four and six, without spores.

Properties.—Very motile; liquefy gelatine rapidly; a peculiar sweetish odor is produced in the cultures, and a blue pigment.

Growth.—Develops readily at ordinary temperature, growing quickly and mostly on the surface; it is ærobic. *Colonies on gelatine plate,* in two or three days a greenish iridescence appears over the whole plate, the colonies having a funnel-shaped liquefaction, and appearing under low power when still young, as yellowish green, the periphery being granulated.

Stab Cultures.—Mainly in upper strata, the liquefaction funnel-shaped, the growth gradually settling at the bottom, a rich green shimmer forming on the surface, and the gelatine having a deep fluorescence.

Potato.—The potato is soaked with the pigment, a deep fold of green occurring on the surface.

Staining.—With ordinary aniline dyes.

Pathogenesis.—When animals are injected with fresh cultures in the peritoneal cavities or cellular tissues, a rapidly spreading œdema with general suppuration develops. The bacilli are then found in the viscera and blood.

If a small quantity is injected, a local suppuration occurs, and if the animal does not die it then can withstand large quantities. It is immune.

The Pigment. Pyocyanin.—When the pus, bandages, and dressings containing the bacillus pyocyaneus are washed in chloroform, the pigment is dissolved and crystallizes from the chloroform in long needles. It is soluble in acidulated water, which is turned red thereby, and when neutralized the blue color returns. It has no pathogenic action. It is an aromatic compound. The bacillus has no especial action on the wound, and is found sometimes in perspiration of healthy persons.

Bacillus Pyocyaneus. β. (Ernst.) A bacillus found in grayish pus-colored bandages.

The only especial difference between this and the above is the formation of brownish-yellow pigment instead of *pyocyanin*. The form and appearance of cultures otherwise the same.

Micrococcus Gonorrhœa. *Gonococcus.* (Neisser.) In 1879 Neisser demonstrated the presence of this germ in the secretion of specific urethritis.

Form.—Cocci, somewhat triangular in form, found nearly always in pairs, the base of one coccus facing the base of the other, and giving the appearance of a Vienna roll, hence the German name Semmel (roll)-form. Four to twelve such pairs are often found together.

FIG. 70.

Gonococci in gonorrhœal pus. Aniline, methyl violet. (650 *diameters.*)

Properties.—No movement of their own.

Culture.—On gelatine-agar or potato they do not grow, and only upon human-blood serum have they given any semblance of a growth. The temperature must be between

33° and 37° C., and the growth occurs very slowly and sparsely.

In three days a very thin, almost invisible, moist yellowish growth, seeming to be composed of little drops.

Under low power small processes are seen shooting out from the smooth border.

FIG. 71.

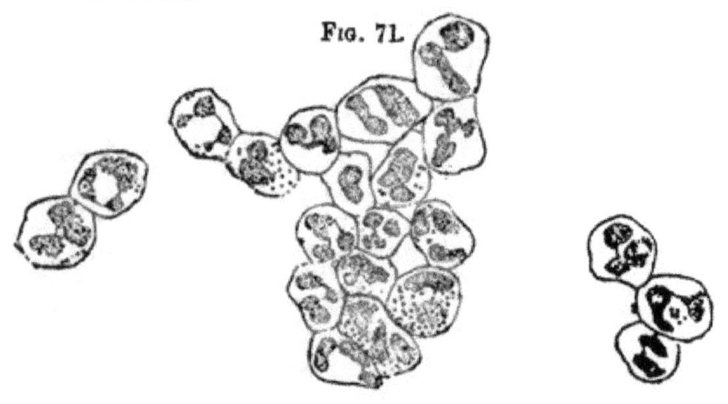

Gonococci in pus.

It requires to be then transferred to fresh media, as it quickly perishes.

Staining.—Colored easily with all ordinary aniline stains.

Gram's method is not applicable, this being one of its main diagnostic features.

The following method for coloring cover-glasses is recommended by Neisser.

The cover-glasses, with some of the urethral discharge smeared upon them, are covered with a few drops of alcoholic solution of eosin and heated for a few minutes over the flame. The excess of the dye is removed with filter paper, then the cover-glass placed in concentrated methylin blue (alcoholic solution) for 15 seconds, and rinsed in water.

The gonococci are dark blue, the protoplasm of the cell pink, and the nucleus a light blue, the gonococci lying in the protoplasm next to the nucleus.

Other bacteria are similar to the gonococci in form ; they are distinguished from the gonococcus, in that they are colored with Gram's method, whereas the micrococcus of gonorrhœa is not. Therefore it is always necessary, after having first found these peculiar-shaped microbes, to apply *Gram's stain*, and if they are then *not found* one can safely say it is the gonococcus.

Pathogenesis.—The attempts to infect the experiment animals with gonorrhœa have so far been without success. In man, upon a healthy urethra, a specific urethritis was produced with even the 20th generation of the culture. *Gonorrhœal ophthalmia* contains the cocci in great numbers, and gonorrhœal rheumatism is said to be caused by the lodgment of the cocci in the joints.

The microbes have been found long after the acute attack, when only a very slight oozing remained, and the same were very virulent.

The specific inflammations of the generative organs of the female are due to this microbe extending its influence, having gained entrance through the vagina. It is found chiefly in the superficial layers of the mucous membrane.

Similar Microbes found in the Urethra and Vagina.

Micrococcus Citreus Conglomeratus. (Bumm.) Very similar to the gonococci in form, they are, however, *easily cultivated*, and form yellow colonies which dissolve the gelatine and grow quite rapidly; the surface of the gelatine is at first moist and shiny, but later on wrinkled. They *are colored with Gram's method*, and have *no special pathological action*. They are found in the air and gonorrhœal pus.

Diplococcus Albicans Amplus. (Bumm.) In vaginal secretion. The diplococci are much larger than the gonococci, but similar in form. They are also cultivated upon gelatine plates, grayish-white colonies, which slowly liquefy gelatine. They grow moderately rapid. Stained with Gram's method, and have no pathogenic action.

Diplococcus Albicans Tardissimus. (Bumm.)

Origin.—In urethral pus.

Form.—Like gonococci.

Properties.—Immotile; do not liquefy gelatine.

Growth.—Very slow at ordinary temperature, but more rapid

at brood-heat. The colonies are as small white points, which under low power appear brown and opaque.

Agar Stroke Culture.—Grayish-white growth, which after two months is like a skin upon the surface.

Staining.—Takes Gram's method.

Pathogenesis —None known.

Micrococcus Subflavus. (Bumm.)

Origin.—In lochial discharges, in vagina and urethra of healthy persons.

Form.—As gonococci.

Properties.—Not motile; liquefy gelatine slowly; a yellow-brownish pigment.

Growth.—Grows slowly on all media, forming on gelatine, after two weeks, a moist yellowish surface growth.

Potato.—Small half-moon-shaped colonies which, after three weeks, become light-brown in color, and covering the surface as a skin.

Staining.—Colored with Gram.

Pathogenesis.—Not acting upon the mucous membrane, but when injected in cellular connective tissue, an abscess results which contains myriads of diplococci.

The **gonococcus** is distinguished from all these similar micrococci by *being found usually within the cell protoplasm.*

Secondly.—Not stained with Gram's method.

Thirdly.—Refusing to grow readily upon gelatine.

All the similar bacteria being easily cultivated.

These characteristics, taken *in toto*, form sufficient features for its ready recognition, and as it is often a serious question to decide, not so much because of the patient's health as because of his character, we should be very careful not to pronounce a verdict until we have tested the micro-organism as above. When the germ so tested is found, the process can be called *specific* without a *doubt*.

Bacillus of Tetanus. (Nicolaier-Kitasato.)

Origin.—Nicolaier found this bacillus in the pus of a wound in one who had died of tetanus, describing it in 1884.

Kitasato has since then been able to isolate and cultivate this germ. (1889.)

Form.—A very delicate, slender rod, somewhat longer than the bacillus of mouse septicæmia, which is the smallest bacillus. When the spores form, a small swelling occurs at the end where the spore lies, giving it a drum-stick shape.

FIG. 72.

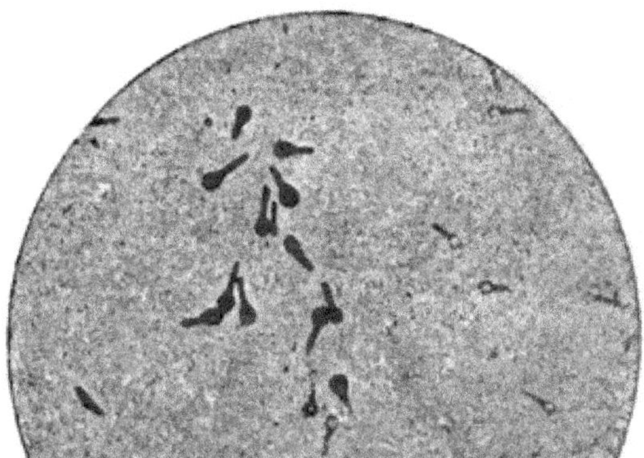

Bacillus of Tetanus with spores.

Properties.—Not very motile, though distinctly so; liquefies gelatine slowly. The cultures give rise to a foul-smelling gas.

Growth.—Develops very slowly, best at brood-heat (36° to 38° C.), and only when all oxygen is excluded, an *obligatory anærobin.* In an atmosphere of carbon dioxide gas it cannot grow, but in hydrogen it flourishes.

Colonies on gelatine plates in an atmosphere of hydrogen. Small colonies. After four days a thick centre and radiating wreath-like periphery, like the colonies of bacillus subtilis.

High Stab-Culture.—(The gelatine having 2 per cent. glucose added and filling the tube.) Along the lower portion of the needle-track, a thorny-like growth, little needle-like points shooting out from a straight line. The whole tube becomes clouded as the gelatine liquefies, and then the growth settles at the bottom of the tube.

Fig. 73.

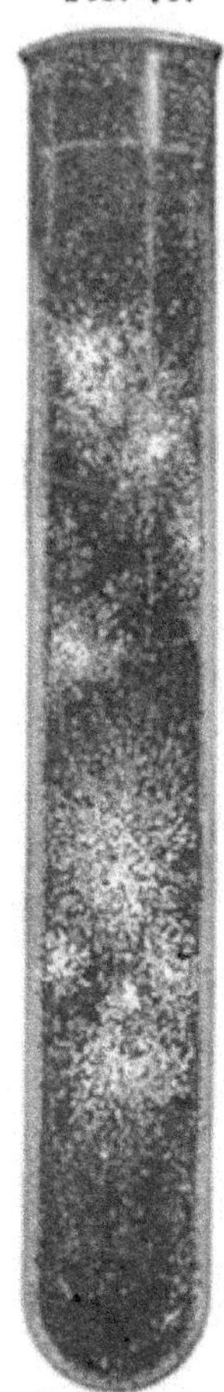

Appearance of culture of bacillus of tetanus after agitating the liquefied gelatine. (Fränkel and Pfeiffer.)

Fig. 74.

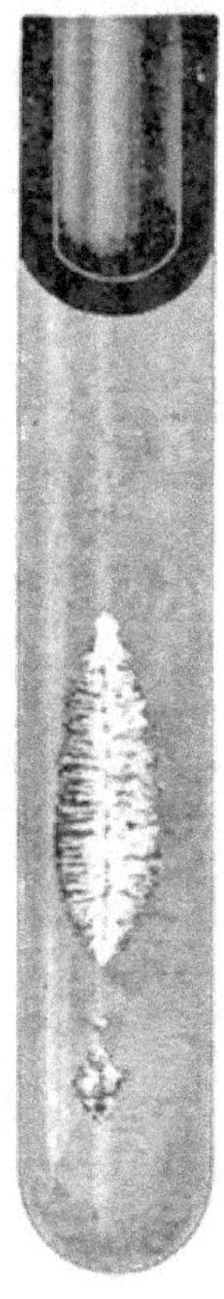

Six days' culture of bacillus of tetanus in gelatine (deep stab). (Fränkel and Pfeiffer.

Agar.—At brood-heat, on agar, the growth is quite rapid, and at the end of forty-eight hours gas bubbles have formed and the growth nearly reached the surface.

Bouillon.—Adding glucose to the bouillon gives a medium in which an abundant growth occurs.

Staining.—All the ordinary stains, Gram's method also; the spores being colored in the usual way.

Pathogenesis.—A small amount of the pure culture injected under the skin of experiment animals will cause, in two to three days, death from true tetanus, the tetanic condition starting from the point of infection. At the autopsy *nothing* characteristic or abnormal is found, and the bacilli have disappeared, except near the point of entrance. This fact is explained as follows:

Several toxic products have been obtained from the cultures, and they are produced in the body, and give rise to the morbid symptoms. These have been isolated, and when injected singly cause some of the tetanic symptoms. The virus enters the circulation, but does not remain in the tissues.

Four ptomaïnes among them: tetanin, tetanotoxin, and spasmotoxin; also a toxalbumen.

Immunity.—Kitasato, by inoculation of sterilized cultures, has been able to cause immunity from the effects of virulent bacilli.

An anti-toxin obtained by Tizzoni and Cattani from the serum of animals made immune by sterilized cultures has been used with curative effects in several cases of tetanus in man. It is a globulin, but differs from the anthrax anti-toxin, and it is found exclusively in the serum. By precipitation with alcohol and drying *in vacuo*, the anti-toxin is obtained in a solid state. The aqueous solution is used for injection subcutaneously.

Habitat.—The bacillus is present in garden earth, in manure; and even from mortar it has been isolated.

The earth of special districts seems to contain the bacilli in greater quantities than in others.

Bacillus Œdematis Maligni. (Koch.)

Vibrion Septique.—(Pasteur.)

Origin.—In garden earth, found lately also in man, in severe wounds when gangrene with œdema had developed. Identical with the bacillus found in Pasteur's septicæmia.

Form.—Rods somewhat smaller than the anthrax bacilli, the ends rounded very sharply. Long threads are formed. Very large spores which cause the rods to become spindle- or drum-stick-shaped.

Properties.—Very motile; liquefy gelatine; do not produce any foul gaseous products in the body.

Growth.—Grows rapidly, but only when the air is excluded, and best at brood or body heat.

Roll Cultures.—(After Esmarch's method.) Small, round glancing colonies with fluid contents, under low power, a mass of motile threads in the centre, and at the edges a wreath-like border.

High Stab Culture.—With glucose gelatine, the growth at first seen in the bottom of the tube, with a general liquefaction of the gelatine, gases develop and a somewhat unpleasant odor.

Agar.—The gases develop more strongly in this medium, and the odor is more prominent.

Guinea-Pig Bouillon.—In an atmosphere of hydrogen clouding of the entire culture medium without any flocculent precipitate until third day.

Staining.—Are stained with the ordinary dyes, but Gram's method is not applicable.

Pathogenesis.—When experiment animals, mice or guinea-pigs, are injected with a pure culture under the skin they die in 8 to 15 hours, and the following picture presents itself at the autopsy: In guinea-pigs from the point of infection, spreading over a large area, an œdema of the subcutaneous tissues and muscles, which are covered and saturated with a clear red serous exudate free from smell. This contains great quantities of bacilli.

The spleen is enlarged, especially in mice. The bacilli are not found in the viscera, but are present in great numbers on the surface, *i. e.*, in the serous coverings of the different organs; though when any length of time has elapsed between the death of the animal and the examination, they can be found in the inner portions of the organs, for they grow well upon the dead body. In man they have been found in rapidly spreading gangrene. They are present in the soil, in putrefactions of various kinds, and in dirty water.

Immunity.—Is produced by injection of the sterilized cultures, and also the filtered bloody serum of animals dead with the disease.

Spirillum of Relapsing Fever. (Obermäier.)

FIG. 75.

Cultures in agar of malignant Œdema, after 24 hours, at 37° C. (Fränkel and Pfeiffer.)

FIG. 76.

Culture in gelatine of malignant Œdema. (Fränkel and Pfeiffer.)

Syn. Spirochæte Obermäieri.

Origin.—Found in the blood of recurrent fever patients, described in 1873.

Form.—Long, wavy threads (16 to 40 μ long), a true spirillum; flagella are present.

Properties.—Very motile. *Has not been cultivated.*

Staining.—Ordinary aniline stains. Bismark brown best for tissue sections.

Pathogenesis.—Found in the organs and blood of recurrent fever. Man and monkeys inoculated with blood from one suffering from this disease become attacked with the fever, and in their blood the spirillum is again found. It is found in the blood, only in the relapses (during the fever). After the attack the spirilla gather in the spleen and gradually die there. It has been found in the brain, spleen, liver, and kidneys. In the secretions it has not been discovered.

Bacillus Malariæ. (Klebs and Tommaci-Crudeli.)

Origin.—These two observers have found a germ present in malarial persons in the blood, which produced an intermittent fever in animals which had been inoculated with such blood. They were also found in the soil of the Roman Campagna. Very little importance is at present attached to this germ, but at the time of its discovery, 1879, it was thought to be the cause of malaria.

Hæmatozoa of Malaria. Certain micro-organisms are found in the blood of persons suffering from malaria, and have lately been very carefully studied. They do not belong to bacteria, being really of animal origin, among the protozoa; but because they are described in the larger works on bacteria, it is necessary that they be considered here.

Synonyms. Hæmatomonas Malariæ (Osler). *Plasmodium Malariæ* (Laveran).

Form.—Various shapes have been described, and whether they are all of one micro-organism or several distinct organisms s not yet definitely settled. They have been found in the cell *intra-corpuscular*, and outside of the cell *extra-corpuscular.*

1st. *Round bodies* 1 to 10 μ. in diameter, lying free in the serum or hanging on to the blood-plates. They have very lively amœboid movements.

2d. *Flagellated bodies.*—Flagella on the upper surface of the full-grown round bodies. These vary in length and are very mobile. They then loosen themselves, and are finally lost. Found only in freshly drawn blood.

3d. *Crescent-shaped bodies.*—These are cylindrical and usually bent sickle-shape. In the centre a black spot formed of little granules. Not motile.

4th. *Rosette bodies.*—Regularly outlined with a collection of pigment-granules in the centre. The rosette afterwards divides and separates into segments.

5th. White blood-corpuscles containing melanin, which they have digested.

FIG. 77.

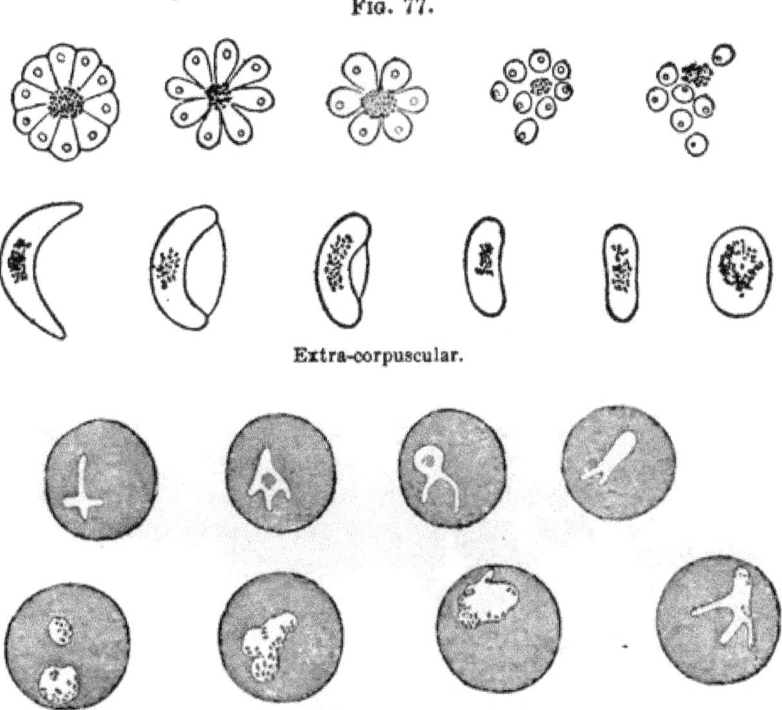

Extra-corpuscular.

Intra-corpuscular.

VARIOUS FORMS OF PLASMODIA.

Cultivation of these organisms has not yet been attained.

Staining and Examination of Blood.—Take the blood of a per son subject to malarial fevers, just before a paroxysm. Having first carefully cleansed the finger, a ligature is applied, and the drop of blood drawn with a needle, brought on a well-cleaned cover-glass, and immediately covered with a second cover-glass. This is now examined with a strong objective (dry system) by day-light.

If, now, a stained preparation is wanted, the cover-glasses are slid apart, passed three times through the flame, and a concentrated solution of methylin-blue left on for a few minutes.

Still better is it to allow a drop of methylin-blue solution in a little ascitic fluid to flow slowly on the cover-glass before the blood has become dry. The finer structure will then be more plainly brought out. Laveran recommends the strong objective of the dry system for examining.

A drop of a watery solution of fuchsin or methylene-blue can be placed on the glass slide and a drop of blood on the cover-glass, then the cover-glass turned over on to the slide so that the two liquids mix, and examined at once.

Pathogenesis.—These organisms have been found only in malarial diseases, and they have been constantly found. Malarial paroxysms have been produced in a healthy person by inoculation of blood containing such organisms. They disappear under the use of quinine.

Golgi finds certain types constant in tertian, and others again peculiar to quartan.

Some, however, hold all these various forms as nothing more than changed blood-corpuscles. The impossibility of obtaining a pure culture leaves the question still in doubt.

Grassi and *Feletti* claim to have produced in sparrows and man, by injection of the blood of malarial persons, malarial fever, and found the specific parasites for the different forms. The amœboid or intra-corpuscular give rise to the typical intermittent fevers. The crescent-shaped extra-corpuscular, producing the dumb ague or irregular fever; four different amœbæ were found.

Hæmamœba	præcox	produces the	quotidian.
"	vivax	" "	tertians.
"	malariæ	" "	quartans.
"	immaculata	" "	quotidian.
Laverania	malariæ	" "	irregular fever.

They place them with the Rhizopoda.

CHAPTER IV.

BACTERIA PATHOGENIC FOR ANIMALS BUT NOT FOR MAN.

Bacillus of Symptomatic Anthrax. (Bollinger and Feser.) (*Charbon symptomatique.* Arloing, Cornevin, and Thomas.)

Origin.—This bacillus, described already in 1879, has only lately been isolated, and by animal inoculation shown to be the cause of the "black-leg" or "quarter-evil" disease of cattle.

Form.—Large slender rods, which swell up at one end or in the middle for the spore.

Properties.—They are motile, and liquefy gelatine quite rapidly.

A rancid odor is developed in the cultures.

Cultures.—The growth occurs slowly, and only in an atmosphere of hydrogen, being very easily destroyed by oxygen and carbon dioxide; grows best at blood heat; under 15° C. no growth.

Glucose-gelatine.—In a few days little round colonies develop, which, under low power, show hairy processes around a compact centre.

Stab Cultures in full test tubes.—The growth first in the lower portion of the tube not very characteristic. Gases develop after a few days, and the gelatine becomes liquid.

Agar at brood temperature, in 24 to 48 hours, an abundant growth with a sour odor and abundant gas formation.

Staining.—Ordinary methods. Gram's method is not applicable to the *rods;* but the spores can be colored by the regular double stain for spores.

Pathogenesis.—If a small amount of the culture be injected under the skin of a guinea-pig, in twenty hours a rise of temperature, pain at the site of injection, and in a few hours more death. At the autopsy, the tissues blackened in color and soaked with a bloody serous fluid; in the connective tissue large collections of gas, but only in the neighborhood of the point of infection. The bacilli are found in great numbers in the serum, but only appear in the viscera some time after death, when spores have developed.

The animals are usually infected through wounds on the extremities; the stalls or meadows having been dirtied by the spore-containing blood of animals previously dead of the disease. "*Rauschbrand*" is the German name; "*Charbon symptomatique*," the French, from the resemblance in its symptoms to anthrax.

Immunity.—Rabbits, dogs, pigs, and fowls are immune by nature, but if the bacilli are placed in a 20 per cent. solution of lactic acid, and the mixture injected, the *disease* develops in them. The lactic acid is supposed to destroy some of the natural resistance of the animal's cells.

When a bouillon culture is allowed to stand a few days, the bacilli therein lose their virulence, and animals are no longer affected by them.

But if they are placed in 20 *per cent. lactic acid* and the mixture injected, their virulence returns.

Immunity is produced by the injections of these weakened cultures, and also by some of the products which have been obtained from the cultures.

Bacillus of Chicken Cholera. (Pasteur.)

Syn.—*Micrococcus cholera gallinarum. Microbe en huit. Bacillus avicidus. Bacillus of fowl septicæmia.*

Fig. 78.

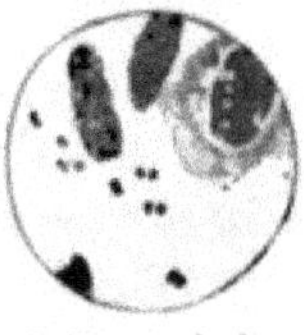

Chicken cholera in blood 1000 ×. (Fränkel and Pfeiffer.)

Origin.—In 1879 Perroncito observed this cocci-like bacillus in diseases of chickens, and Pasteur, in 1880, isolated and reproduced the disease with the microbe in question.

Form.—At first it was thought to be a micrococcus, but it has been seen to be a short rod about twice as long as it is broad, the ends slightly rounded. The centre is very slightly influenced by the aniline colors, the poles easily, so that in stained specimens the bacillus looks like a dumb-bell or a figure-of-eight. (Microbe en huit.)

Properties.—They do not possess self-movement; do not liquefy gelatine.

Growth.—Occurs at ordinary temperature, requiring oxygen for development. It grows very slowly.

Gelatine Plates.—In the course of three days little round, white colonies, which seldom increase in size, having a rough border and very finely granulated.

Stab Culture.—A very delicate gray line along the needle-track, which does not become much larger.

Agar Stroke Culture.—A moist, grayish-colored skin, more appreciable at brood heat.

Potato.—At brood heat after several days a very thin, transparent growth.

Staining.—Methylin blue gives the best picture. Gram's method is not *applicable*. As the bacillus is easily decolorized, aniline oil is used for dehydrating tissue sections, instead of alcohol.

Method:

Löffler's methylin blue	½ hour.
Alcohol	5 seconds.
Aniline oil	5 minutes.
Turpentine	1 minute.
Xylol and Canada balsam.	

Pathogenesis.—Feeding the fowls or injecting under the skin will cause their death in from 12 to 24 hours, the symptoms preceding death being those of a heavy septicæmia.

The bacillus is then found in the blood and viscera, and the intestinal discharges, the intestines presenting a hemorrhagic inflammation.

Guinea-pigs and sheep do not react. Mice and rabbits are affected in the same manner as the fowls.

Immunity.—Pasteur, by injecting different-aged cultures into fowls, produced in them only a local inflammation, and they were then immune. But as the strength of these cultures could not be estimated, many fowls died and the healthy ones were endangered from the intestinal excretions, which is the chief manner of infection naturally ; the fæces becoming mixed with the food.

Bacteria of Hemorrhagic Septicæmia. (Hueppe.)

Under this heading Hueppe has gathered a number of bacteria very similar to the bacillus of chicken cholera, differing from it and each other but very little. They have been described by various observers and found in different diseases.

(1) The bacteria of this group color themselves strongly at the *poles*, giving rise to the dumb-bell shape. They do not take the *Gram stain.* They are without spores,

(2) And do not liquefy gelatine.

They have been placed in three general divisions:—

1st division.
- Wild Plague. (Hueppe.)
- German Swine Plague. (Löffler, Schütz.)
- Rabbit Septicæmia.
- Ox Plague. (Oresti-Armanni.)
- Steer Plague. (Kitt.)

The bacteria of the first division are not motile, do not grow on potato, and are found scattered through the bloodvessels. A local reaction is uncommon.

2d division.
- American Swine Plague. (Billings.)
- French Swine Plague. (Cornil and Chantemesse.)
- Cattle Plague. *Texas Fever.* (Billings.)
- Frog Plague. (Eberth.)

Here the bacteria are *motile.* They grow on potatoes and are similar to the typhoid bacillus in gelatine. They form small embolic processes in the capillaries. They cause only a local disturbance in rabbits when subcutaneously injected. An acid fermentation is produced in milk.

3d division.
- Hog Cholera. (Salmon.)
- Swedish Swine Plague. (Lelander.)

The bacteria of this third division are very motile. The hog-cholera bacilli lie in the spleen and other organs in small masses like the typhoid bacillus.

Rabbits die in four to eight days without any local disturbance. The growth on potato is strong.

The Swedish swine-plague bacillus occupies a position between that of *Hog Cholera* and *Bacillus Coli Communis.*

The various swine-plague bacilli are but little active in fowls, differing thus widely from the chicken-cholera bacillus.

Bacillus of Erysipelas of Swine. (Löffler, Schütz.) *Schweinerotlaufbacillus* (*German*). *Rouget du porc* (*French*).

Origin.—Found in the spleen of an erysipelatous swine by Löffler in 1885.

Form.—One of the smallest forms of bacilli known ; very thin, seldom longer than 1 μ, looking at first like little needle-like crystals. Spores have not been found.

Properties.—They are motile ; do not liquefy gelatine.

Growth in culture at ordinary temperature. very slowly, and the less oxygen the better the growth.

Gelatine Plate.—On third day little silver-gray specks, seen best with a dark background, coalescing after awhile, producing a clouding of the entire plate.

Stab Cultures.—In a few days a very light, silvery-like clouding, which gradually involves the entire gelatine ; held up against a dark object, it comes plainly into view.

Staining.—All ordinary dyes and Gram's method also.

Tissue sections stained by Gram's method show the bacilli in the cells, capillaries, and arterioles in great numbers.

Pathogenesis.—Swine, mice, rabbits, and pigeons are susceptible ; guinea-pigs and chickens, immune.

When swine are infected through food or by injection a torpidity develops with diarrhœa and fever, and on the belly and breast red spots occur which coalesce, but do not give rise to any pain or swelling. The animal dies from exhaustion in 24 to 48 hours. In mice the lids are glued together with pus.

At the autopsy the liver, spleen, and glands are enlarged and congested, little hemorrhages occurring in the intestinal mucous membrane and that of the stomach.

Bacilli are found in the blood and all the viscera.

One attack, if withstood, protects against succeeding ones.

Immunity.—Has also been attained by injecting vaccines of two separate strengths.

Bacillus Murisepticus. (Koch.) *Mouse septicæmia.*

Origin.—Found in the body of a mouse which had died from injection of putrid blood, and described by Koch in 1878.

Form.—Differs in no particular from the bacillus of swine erysipelas, excepting that it is a very little shorter, making it the *smallest* known bacillus. Spores have been found, the cultures exactly similar to those of swine erysipelas.

The pathological actions are also similar. Field mice are immune; whereas for house and white mice the bacillus is fatal in two to three days.

Micrococcus of Mal de Pis. (Nocard.) Gangrenous mastitis of sheep.

Origin.—In the milk and serum of a sheep sick with the "*mal de pis.*"

Form.—Very small cocci seldom in chains.

Properties, immotile; liquefying gelatine.

Growth.—Growth occurs best between 20° and 37° C., is very rapid, and irrespective of oxygen.

Plates of Gelatine.—White round colonies, some on the surface and some in the deeper strata, with low *power*, appearing brown surrounded by a transparent areola.

Stab Culture.—Very profuse along the needle-track, in the form of a cone after two days, the colonies having gathered at the apex.

Potato.—A dirty gray, not very abundant, layer somewhat viscid.

Staining, with ordinary methods; also Gram's method.

Pathogenesis.—If a pure culture is injected into the mammary gland of sheep, a "*mal de pis*" is produced which causes the death of the animal in 24 to 48 hours. The breast is found œdematous, likewise the thighs and perineum; the mammæ very much enlarged, and at the nipples a blue-violet coloration. The spleen is small and black; other animals are less susceptible. In rabbits abscesses at the point of infection, but no general affection.

Bacillus Alvei. (Cheshire and Cheyne.) *Bacillus melittophtharus.* (Cohn.)

Origin.—In foul-brood of bees.

Form.—Slender rods, with round and conical-pointed ends; very large oval spores, the rod becoming spindle-shaped when they appear.

Properties.—Motile, liquefying gelatine rapidly.

Growth.—Grows best between 20° C. and 37° C., very slowly; ærobic.

Gelatine Plates.—Small grooves are slowly formed, which unite so as to form a circle or pear-shaped growth, from which linear grooves again start.

Stab Culture.—Grows first on surface, then gradually along the needle-track, long processes shooting out from the same, clouding the gelatine. Later, air-bubbles form like the cholera culture, and in two weeks the whole gelatine liquefied.

Staining.—Do not take aniline dyes very well. Gram's method is, however, applicable.

Pathogenesis.—If a pure culture is spread over the honey-comb containing bee larvæ, or if bees are fed upon infected material, foul-brood disease will occur. Mice, if injected, die in a few hours. Œdema around the point of infection, and many bacilli contained in the œdematous fluid, otherwise no changes.

Micrococcus Amylivorus. (Burrill.)

Origin.—In the disease called "*Blight,*" which affects pear-trees and other plants.

Form.—Small oval cells, never in chains, more the form of a bacillus.

Pathogenesis.—Introduced into small incisions in the bark of pear-trees the trees perished from the "blight." The starch of the plant cell was converted into carbon dioxide, hydrogen, and butyric acid.

Bacterium Termo. (Cohn.)

This was a name given to a form of micro-organism found in decomposing albuminous material, and was supposed to be one specific germ. Hauser, in 1885, found three different distinct microbes which he grouped under the common name of Proteus, which have the putrefying properties ascribed to B. Termo.

Proteus Vulgaris.

Origin.—In putrid animal matter, in meconium and in water.

Form.—Small rods, slightly curved, of varying lengths, often in twisted chains, having long cilia or flagella.

Properties.—Very motile, and very soon liquefying gelatine; forms hydrogen sulphide gas; causes putrefaction in meat.

Growth.—Growth very rapid, best at 24° C., is facultative ærobic.

Gelatine Plates.—Yellowish-brown, irregular colonies, with prolongations in every direction, forming all sorts of figures; an impression preparation shows these spider-leg processes to consist of bacilli in regular order.

Stab Culture.—The gelatine soon liquid, a gray layer on the surface, but the chief part of the culture in small crumbs at the bottom.

Pathogenesis.—Rabbits and guinea-pigs injected subcutaneously die quickly, a form of toxæmia, hemorrhagic condition of lungs and intestines present. When *neurin* is injected previously the animals do not die. This ptomaïne is supposed to be generated by the proteus vulgaris.

Proteus Mirabilis. (Hauser.)

Differs from P. vulgaris in that the gelatine is less rapidly liquefied. Found also in putrid material.

Proteus Zenkeri. (Hauser.)

Does not liquefy gelatine; otherwise similar to the other two.

We have now considered some of the characteristics of the more important bacteria. The scope of this work does not allow a more extended study than we have made, which, as we are aware, has been very superficial. The larger works must be referred to, if a deeper interest is taken in the subject.

APPENDIX.

YEASTS AND MOULDS.

In works on bacteria, these true fungi, *yeasts and moulds*, are usually considered. They are so closely related to bacteria, and so often contaminate the culture media, and are so similar in many respects, that a description is almost a necessity.

But there are several thousand varieties, and we cannot attempt to describe even all of the more important ones. It will answer our purpose to detail a few of the more common kinds, and give the principal features of the different orders.

Fungi exist without chlorophyl.

Saccharomycetes or *Yeasts* increase through budding; the spores attached to the mother cell like a tuber on a potato.

Yeasts are the cause of alcoholic fermentation in the saccharoses. A description of the most common ones will suffice.

Saccharomyces Cerevisiae. (*Torula Cerevisiæ.*) This is the ordinary beer yeast.

Form.—Round and oval cells; a thin membrane inclosing a granular mass, in which usually can be seen three or four irregular-shaped spores. When these become full grown they pass through the cell wall and form a *daughter* cell. Sometimes long chains are produced by the attached daughter cells.

Growth.—They can be cultivated as bacteria in bouillon, but they grow best in beer.

There are several varieties of beer yeast, each one giving a characteristic taste to the beer. Brewers, by paying special attention to the nutrient media, cultivate yeasts which give to their beers individual flavors.

Mixed yeast gives rise to a poor quality of beer.

Saccharomyces Rosaceus. S. Niger and S. Albicans. These yeasts are found in the air; and instead of producing alcoholic fermentation they give rise to a pigment in the culture media. They grow upon gelatine which they do not liquefy.

Saccharomyces Mycoderma. This yeast forms a mould-like growth, a skin, on the surface of fermented liquids, but does not cause any fermentation itself. It forms the common "mould" on wine, preserves, and "sour krout."

Oidium. A form which seems to be the bridge between the yeast and the moulds is the oidium. Sometimes it resembles the yeasts, sometimes the moulds, and often both forms are found in the same culture. Several are pathogenic for man.

Oidium Lactis.

Origin.—In sour milk and butter.

Form.—The branches or hyphens break up into short rod-like spores. No sporangium, as in moulds.

Growth.—In milk it appears as a white mould.

Artificially cultured on gelatine plates, or milk gelatine plates, it forms satin-like, star-shaped colonies, which slowly liquefy. Under microscope the form of the fungus is well seen.

Agar Stroke Culture.—The little stars, very nicely seen at first; then the culture becomes covered with them, causing a smeared layer to appear over the whole surface, with a sour odor.

Properties.—The milk is not changed in any special way. It is not pathogenic for man or animals. It is found when the milk begins to sour.

Oidium Albicans. (*Soor.*) *Thrush Fungus.*

Origin.—Mucous membrane of the mouth, especially of infants.

Form.—Taken from the surface of the culture, a form like yeasts; but in the deeper layers, mycelia with hyphens occurs.

Growth.—Not liquefying; snow-white colonies on gelatine plates.

Stab Culture.—Radiating yellow or white processes spring from the line made by the needle, those near the surface having oval ends.

Potatoes.—The yeast form, develops as thick white colonies.

Bread Mash.—Snow-white veil over the surface.

Pathogenesis.—In man the parasitic thrush, or "white mouth," is caused by this fungus. In the white patches the spores and filaments of this microbe can be found. Rabbits receiving an intravenous injection perish in twenty-four to forty-eight hours, the viscera being filled with mycelia.

True Moulds. Flügge has made five distinct divisions of moulds. It will, however, serve our purpose to classify those to be described under three headings: *Penicillium*, *Mucor*, and *Aspergillus*.

Penicillium Glaucum.

Origin.—The most widely distributed of all moulds, found wherever moulds can exist.

Form.—From the mycelium, hyphens spring which divide into basidia (branches), from which tiny filaments arise (sterygmata), arranged like a brush or tuft. On each sterygma a little bead or conidium forms, which is the spore. In this particular fungus the spores in mass appear green.

Growth.—It develops only at ordinary temperatures, forming thick grayish-green moulds on bread-mash. At first these appear white, but as soon as the spores form, the green predominates. Gelatine is liquefied by it.

Mucor Mucedo. Next to the penicillium glaucum, this is the most common mould. Found in horse dung, in nuts, and apples, in bread and potatoes as a white mould.

Form.—The mycelium sends out several branches, on one of which a pointed stem is formed which enlarges to form a globular head, a spore-bulb, or *Sporangium*. The spore-bulb is partitioned off into cells in which large oval spores lie. When the spores are ripe a cap forms around the bulb, the walls break down and the wind scatters the spores, leaving the cap or "*calumella*" behind.

Growth.—Takes place at higher temperatures on acid media. *It is not Pathogenic.*

Achorion Schonleinii.

Tricophyton Tonsurans.

Microsporon Furfur.

These three forms are similar to each other in nearly every particular and resemble in some respects the oidium lactis, in other ways the mucors. The first one, *Achorion Schönleinii*, was discovered by Schönlein in 1839, in *Favus*, and is now known as the direct cause of this skin disease.

Origin.—Found in the scaly crusts of *favus.*

Form.—Similar to oidium lactis.

Growth.—Is very sparse. On gelatine round white masses inclosed by a zone of liquefied gelatine.

In milk it is destroyed.

Pathogenesis.—Causes favus in man.

Tricophyton Tonsurans. Found, in 1854, by Bazin, in Tinea.

Form.—Similar to the achorion or favus fungus.

Growth.—Somewhat more rapid than the favus, and the gelatine quickly liquefied. Old cultures are of an orange-yellow color. Colonies have a star-shaped form.

Pathogenesis.—Herpes tonsurans and the various tineæ are produced by this fungus.

Microsporon Furfur. Found in tinea versicolor, almost identical with the above, forms dry yellow spots, usually on the chest in persons suffering from wasting diseases.

Aspergillus Glaucus.

Origin.—In saccharine fruits.

Form.—The hyphen has formed upon its further end a bulb, from which pear-shaped sterygmata arise and bear upon their ends the conidia or spores.

Growth.—Best upon fruit juices. *Non-pathogenic.* The mould is green. *Aspergillus flavus* has the tufts and spores of a yellow color.

A. Fumigatus. Is pathogenic for rabbits when injected into them. At the autopsy their viscera are found filled with the mould.

Examination of Yeasts and Moulds. Yeasts and moulds are best examined in the unstained condition. A small portion of the colony rubbed up with a mixture of alcohol and a few drops of liquor ammonia; of this, a little is brought upon the glass-slide covered with a drop of glycerine and the cover-glass pressed upon it. If the preparation is to be saved, the cover-glass is secured by sealing-wax around the edges. *Yeasts* take *methyline-blue* stain very well.

Ray Fungus. A division containing the *actinomyces.* (Bollenger and Israel.)

Origin.—In actinomycosis of man and cattle, in the growth.

Form.—In the pus or scrapings, little yellow grains about the size of a pin's head are seen by the naked eye. When one of these points is flattened out between the cover-glass and slide and placed under microscope (200 x), aster-shaped figures will

FIG. 79.

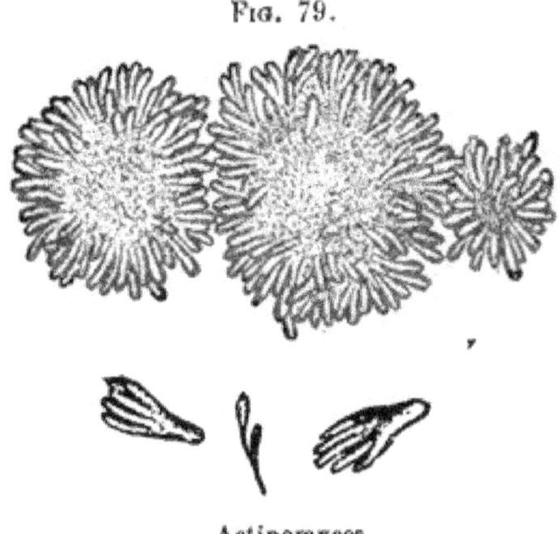

Actinomyces.

be seen, the centre thick, radiating from it, little hyphens, which become thicker and rounder at their peripheral end. These bottle-shaped hyphens are supposed to be the spore-bearing organs. Some of these may have separated from the main body and lie loose or attached to it by a very thin filament.

Growth.—Develops only at brood heat and by exclusion of oxygen.

In Agar.—After several weeks a yellowish growth was obtained, but this consisted mainly of mycelia, the club-shaped or conical rays not forming.

In eggs a growth developed when the method of Hueppe was carried out.

Pathogenesis.—When a portion of the growth obtained in eggs was injected into the abdominal cavity of a rabbit, actinomycotic processes developed upon the peritoneum.

It usually gains access to the living body through a wound in the gum or some caries of the teeth. A new growth is formed, ulceration being first set up.

The new tissue, composed of round cells, then undergoes softening, purulent collections form and the normal structure is destroyed.

The usual seat is in the maxillary bones, but the fungus has been found in the lungs, tonsils, intestines, and various other organs in man and cattle.

Examination.—Well seen in the unstained condition. From the pus or scraping a small portion is taken and squeezed upon the glass slide; if calcareous matter is present, a drop of nitric acid will dissolve the same.

Glycerine will preserve the preparation.

Staining.—Cover-glass specimens stained best with Gram's method. Tissue sections should be stained as follows:—

Ziehl's carbol-fuchsin, ten minutes. Rinse in water.

Conc. alcohol sol. of picric acid, five minutes. Rinse in water.

Alcohol, 50 per cent., fifteen minutes. Alcohol absolute, clove oil, balsam.

The rays stained red, the tissue yellow.

Examination of Air, Soil, and Water.

Air.—Many germs are constantly found in the atmosphere about us. Bacteria unaided do not rise into the air and fly about; they usually become mixed with small particles of dirt or dust and are moved with the wind. The more dust the more bacteria, and therefore the air in summer contains a greater number than the air in winter, and all the other differences can be attributed to the greater or less quantity of dust and wind.

Methods of Examination. The simplest method is to expose a glass or dish covered with gelatine in a dust-laden atmosphere or in the place to be examined. In the course of 24 to 48 hours colonies will be seen formed wherever a germ has fallen. But this method will not give any accurate results in regard to the number of bacteria in a given space; for such a purpose somewhat more complicated methods are needed, so that a certain amount of air can come in contact with the culture media at a certain regulated rate of speed.

Hesse's Method. This is the most useful of the various methods in vogue.

A glass cylinder, 70 centimetres long and 3.5 centimetres in diameter, is covered at one end, by two rubber caps, the inner one having a hole in its centre 10 millimetres in diameter; and at the end *B* a rubber cork fits in the cylinder; through this cork a glass tube 10 mm. in diameter passes, which is plugged at both ends with cotton. The cylinder and fittings are first washed in alcohol and sublimate and then placed for one hour in the steam chamber.

Removing the cork of the cylinder, 50 cubic centimetres of sterile gelatine in a fluid condition are introduced and rolled out on the sides of the tube, after the manner of Esmarch, leaving a somewhat thicker coating along the under side of the

Fig. 80.

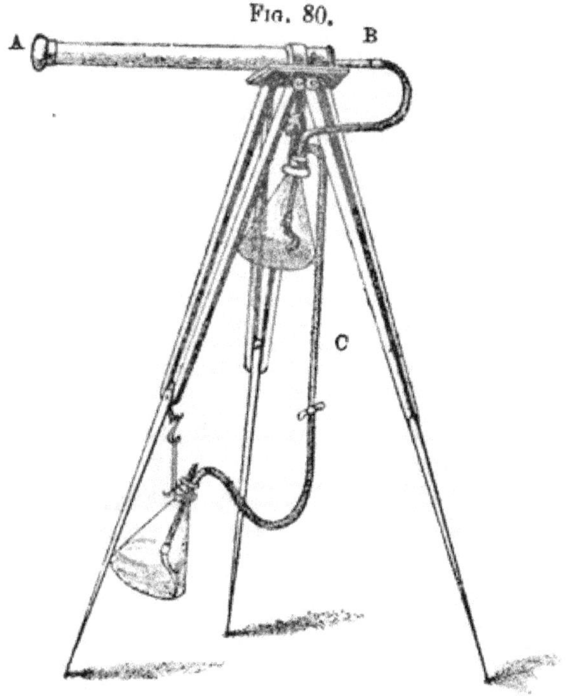

cylinder. The *aeroscope*, as the cylinder and its fittings are called, is placed upon an ordinary photographer's tripod and the glass tube, which passes through the rubber cork, connected with an *aspirator*, the cotton having first been removed from its outer end. The aspirator consists of two ordinary wash-bottles connected with each other by a rubber tube, *C*. They are attached to the tripod with a small hook one above the other, the upper one half filled with water and slightly tilted.

FIG. 81.

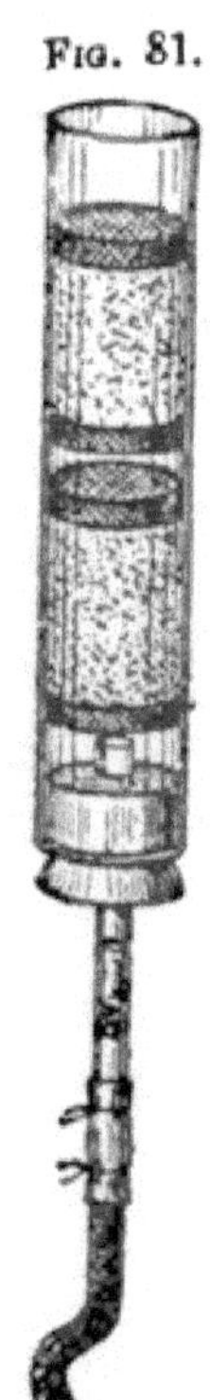

Sand filter after Petri.

When the apparatus is wanted, the outer rubber cap at the end *A* of the aeroscope is removed, the air can then pass through the small hole in the other cap, and the germs fall upon the gelatine in the tube, the cotton in the small glass tube at the other end preventing the germs from getting out. The aspirator is set in use by tilting the upper bottle so that the water flows into the lower, this creates suction and draws the air through the aeroscope.

The amount entering estimated by the capacity of the wash-bottle. The rate at which it enters depending upon the rate of the flow of water, which can be regulated.

Hesse advises for rooms and closed spaces 1 to 5 litres, at the rate of 2 minutes a litre, and for open spaces, 10 to 20 litres at 4 minutes a litre. Plate cultures can be made from the colonies which develop in 8 to 10 days in the cylinder.

Petri's Method. The air pumped or sucked through sand filters, and the sand then mixed with gelatine.

Sand is sterilized by heating to redness, and while still warm placed in test tubes which are then plugged. (Sand which has been passed through a sieve with meshes 0.25 millimetres wide is the kind required.) A glass tube 9 centimetres long is provided with two portions of sand each 3 cm. long and ½ cm. apart, little plates of brass gauze keeping the portions in position.

The tube and its contents now sterilized in hot air oven at 150° C., the ends having first been plugged with cotton.

One end of the tube is then fitted with a rubber cork through which passes a glass tube, which is connected with an aspirator (a hand-pump with a known capacity).

If a hundred litres of air pass through the tube in fifteen minutes the germs should all be arrested in the first sand filter.

And when the filters are removed and thoroughly mixed with gelatine, each filter for itself, there should be no colonies developed from the second filter, *i. e.*, the one nearest the aspirator.

Varieties Found in Air. The only *pathogenic* ones found with any constancy are the staphylo-coccus aureus and citreus; but any bacterium can be, through accident, lifted into the atmosphere, and in certain places may be always found—the bacillus tuberculosis, for example, in rooms where many consumptives are living.

Non-Pathogenic. The micrococci predominate. Sarcina, yeasts, and moulds constantly contaminate cultures.

In the ordinary habitations the average number of germs to the litre of air does not exceed five.

Around water-closets, where one would imagine a great number to exist, owing to the undisturbed condition of the air, but few will be found.

Examination of Water. The bacteriological examination of water is to-day of as much importance as the chemical analysis, and must go hand in hand with it.

At the start we must say that a water containing thousands of germs to the cubic centimeter is far less dangerous than one containing but 2 germs, if one of these two be a typhoid bacillus. It is not the number that proves dangerous; it is the kind.

If a natural water contains more than 500 germs to the cubic centimeter, it were well to examine its source.

Bacteriology performs the *greatest service* in testing the *devices* which are intended to render water fit for *drinking*.

As a diagnostic aid the examination is of but little use. An epidemic of typhoid fever occurs, the water is suspected, an examination is undertaken; but the days of incubation and the days passed before the water is analyzed have given the typhoid germs, if any had been present, ample time to disappear, since in water that contains other bacteria they live a very short time only. Again, the water tested one day may be entirely free and the next day contain a great number, and before the typhoid germ can be proven to be present in that particular water, the epidemic may be past.

Purity of Waters. The purest water we have is the natural spring water—water that has slowly filtered its way through various layers of gravel and sand and comes finally clear and sparkling from the ground. It is without germs; but let such a water stand walled up in cisterns or wells, it becomes as surface water, open to all sorts of impurities, and the bacterial nature of it changes every moment.

Artesian or Driven Well. The *drivrn well* will secure to a certain extent a pure water. It is the only form of well or cistern that will insure this, since the water does not become stagnant in it; but it may connect with an outhouse, the soil being very loose, allowing the products of germs of refuse water to find their way into the well. If a chemical examination shows increased amounts of chloride of sodium, a contamination can be mooted.

Filtered Water. Dangerous as surface water is, the greater quantity used, is such: the inhabitants of larger towns and cities using chiefly the rivers and other large waters which course near them for drinking purposes. A purification or filtration can in a certain measure render these waters harmless.

Filtration is often carried on on a large scale in the waterworks of cities and towns.

Bacteriological examination is here of great service to determine if a water, which has been filtered and may have a very clear appearance, and give no harmful chemical reaction, yet be entirely free, or nearly so, from germs; in other words, if the filter is a germ filter or not.

Charcoal Sponge and Asbestos, the materials formerly in use are objectionable because germs readily develop on them and clog them, so that they require frequent renewal. In very large filters, sand and gravel give the best results; the number of germs in a cubic centimetre is reduced to forty or fifty and kept at that number. This is a very pure water for a city water, though, as we stated before, not a safe one, for among those forty germs very dangerous ones may be found. It is then necessary for the users to refilter the water before drinking it, through a material which will not allow any germs to pass.

Pasteur-Chamberland Filter. This very *perfect filter*, which is now in almost universal use, consists of a piece of polished porcelain in the form of a cylinder closed at one end and pointed at the other. It is placed in another cylinder of glass or rubber and the pointed portion connected with a bottle containing the water, or directly with faucet of the water-pipe. The water courses through the porcelain very slowly and comes out entirely free from germs; pipe-clay, bisque, infusorial earth, and kaolin are also perfect filters. The only disadvantage is the long time it takes for the water to pass through. Pressure is used to accelerate the passage in the form of an aspirator or air-pump. (See Fig. 41.)

The force of the hydrant water is also sufficient to produce a steady, small stream.

These porcelain cylinders can easily be sterilized and the pores washed out.

All the cylinders or bougies are not germ proof, so that they must be tested, and most of them must be cleaned every fourth day, or they will allow germs to pass through.

Boiling as a means of purifying. When such a filter cannot be obtained, the only alternative is to boil all the water to be used for drinking; and this should especially be done in times of typhoid and cholera epidemics.

Methods of Examination. Since the germs rapidly multiply in stagnant water, an examination must not be delayed longer than an hour after the water has been collected. Every precaution must be taken in the way of cleanliness to prevent contamination; sterilized flasks, pipettes, and plugs should, or rather must, be at hand, and the gelatine tubes best inoculated on the spot. If this cannot be done, the sample should be packed in ice until it arrives at the laboratory, which, as before stated, should not be later than an hour after collection. The sample is placed in a sterilized glass flask, and the flask then closed with a sterile cotton plug. A sterilized pipette is then dipped into the flask and 1 c.c. of the water withdrawn in it and added to a tube of gelatine, the gelatine being in a fluid condition. To a second tube, ½ c.c. is added. The tubes are then shaken so as to thoroughly mix the water with the gelatine, and then poured upon wide glass plates—one plate for each tube; the plates are then placed in the moist chamber, and in two to three days examined. If the germs are equally divided, there should be one-half the number on one plate that there is on the other; thus the ½ c.c. serves as control.

Water that is very rich in germs requires dilution with sterilized water 50 to 100 times.

To count the colonies which develop upon the plates, a special apparatus has been designed, for, unaided, the eye cannot see them all.

Wolfhugel's Apparatus. A glass plate divided into squares, each a centimeter large, and some of these subdivided. This plate is placed above the gelatine plate with the colonies, and the number in several quadrants taken, a lens being used to see the smaller ones.

The petri saucers can be used instead of plates, and an apparatus on the Wolfhügel plan can be obtained to count the colonies. It is best to count all the colonies on the plate or dish.

Varieties Found. The usual kinds found are non-pathogenic, but, as is well known, typhoid and cholera are principally spread through drinking water, and many other germs may and do find their way into the water. Many of the common varieties give rise to fluorescence, or produce pigment.

Eisenberg gives 100 different varieties as ordinarily found. As mentioned before, 2 bacteria to a cubic centimeter, one of them typhoid, give more danger to a water than thousands of non-pathogenic ones. When, however, more than 200 bacteria to the c.c. are found, such a water ought not to be considered potable. Distilled water forms often a good medium for some bacteria.

The Examination of the Soil. The upper layers of the soil contain a great many bacteria, but because of the difficulty in analyzing the same, the results are neither accurate nor constant. The principal trouble lies in the mixing of the earth with the nutrient medium; little particles of ground will cling to the walls of the tube, or be imbedded in the gelatine, and may contain within them myriads of bacteria. As with water, the soil must be examined immediately or very soon after it is collected, the bacteria rapidly multiplying in it.

When the deeper layers are to be examined, some precautions must be taken to avoid contamination with the other portions of the soil. One method, very laborious and not often practical, is to dig a hole near the spot to be examined and take the earth from the sides of this excavation.

Frankel's Borer. Fränkel has devised a small apparatus in the form of a borer, which contains near its lower end a small cavity, which can be closed up by turning the handle, or opened by turning in the opposite direction.

It is introduced with the cavity closed, and when it is at the desired depth, the handle is turned, the earth enters the cavity, the handle again turned, incloses it completely, and the borer is then withdrawn.

The earth can then be mixed with the gelatine in a tube, and this gelatine then rolled on the walls of the tube after the manner of Esmarch, or it can be poured upon a glass plate, and the colonies developed so.

Another method is to wash the earth with sterilized water, and the water then mixed with the gelatine, as many of the germs are taken up by the water.

The roll-cultures of Esmarch give the best results, many of the varieties usually found being anærobic.

Animals inoculated with the soil around Berlin die almost always of *maligna nt œdema*, and with that of some other towns invariably of *tetanus*. Many of the germs found are nitrogen formers and play a great rôle in the economy of the soil.

CONCLUSION. In tracing thus briefly the characteristics of the more important bacteria, and the various methods used in studying them, we are conscious of the very superficial manner in which this has been done. We excuse ourselves, however, on the ground that this work is but a wedge with which to enter upon the study, or, for those who do not care to proceed further, an eminence from which a fair view of the ground can be obtained. In this, its humble mission, we trust it may meet with success.

CHIEF CHARACTERISTICS

PART I.—

Name.	Genus.	Biology.	Product.
ACETI.	Bacillus.	Short motile rods in zooglœa; ærobic.	Ferment.
ACIDI LACTICI.	Bacillus.	Short, immotile rods; ærobic.	. . . .
ACIDI LACTICI.	Bacillus.	Short, immotile rods.	. . . . .
ACTINOBACTER.	Bacillus.	Immotile rods with capsule; facul. anærob.	. . . . . .
AEROGENES.	Bacillus.	Small motile rods, single and in pairs; very resistant.	. . . . . .
ÆROPHILUS.	Bacillus.	Slender rods in threads; immotile; oval spores; ærobic.	. . . . . .
AGILIS.	Micrococcus.	Mobile diplococci with fine flagella.	Red pigment.
ALBA.	Beggiatoa.	Cocci and spirals with sulphur.	. . . . . .
ALBA.	Sarcina.	Small cocci in packets.	White pigment.
ALBICANS AMPLUS.	Micrococcus.	Large cocci and diplococci.	. . . . . .
ALBICANS TARDISSIMUS.	Micrococcus.	Diplococci colored by Gram.	. . . . . .
ALBICANS TARDUS.	Micrococcus.	Diplococci not motile.	. . . . . .
ALLII.	Bacillus.	Very small rods.	Alkaloid pigment.
AMYLIFERUM.	Spirillus.	Rigid spirilla with spores; turns blue with iodine.	. . . . . .

OF THE PRINCIPAL BACTERIA.

NON-PATHOGENIC BACTERIA.

Culture Characters.	Actions.	Habitat.	Discoverer.
Not liquefy; membranous growth.	Produces acetic-acid fermentation.	Air.	Kützing.
Not liquefy; small white points porcelain-like; slow.	Lactic-acid fermentation; precipitates caseine.	Air; sour milk.	Pasteur.
Growth faster than above; appearance same.	Alcohol is formed after the lactic-acid fermentation.	Sour milk.	Grotenfeldt.
. . . .	Causes fermentation with gas and alcohol.	Air.	Duclaux.
Rapid growth; round, concentrically-arranged colonies; not liquefy.	. . . . . .	Digestive tract.	Miller.
Liquefy rapidly; small yellow-gray colonies.	. . . . . .	Old cultures.	Liborius.
Slowly liquefying, forming a cone with rose-red color.	. . . . . .	Drinking-water.	Ali Cohens.
. . . .	. . . . . .	Sulphur springs.	Vauch.
Slow growth in small white colonies.	. . . . .	Air and water.	
Slowly liquefy; gray colonies; growth fairly rapid.	Is colored by Gram's method.	Vaginal secretion.	Bumm.
Small white points, not liquefying; very slow growth.	. . . . . .	Urethral pus.	Bumm.
Grows slowly on surface, the boundary raised; twice as large as above.	. . . . . .	Skin in eczema.	Unna, Tommasoli.
Bright green pellicle on agar.	Decomposes albumin.	Green slime of onions.	Griffths.
. . . .	. . . . . .	Water.	VanTiegham.

NON-PATHOGENIC

Name.	Genus.	Biology.	Product.
AMYLOBACTER.	Bacillus.	See *Butrycus*, with whi	ch it is identical.
AQUATILIS.	Micrococcus.	Very small cocci in irregular groups.	. . . . . .
ARACHNOIDEA.	Beggiatoa	Very thick filaments containing sulphur; motile.	. . . . . .
ARBORESCENS.	Bacillus.	Thin rods, with rounded ends in threads, and singly; immotile.	Yellow pigment.
ATTENUATUM.	Spirillum.	Threads with narrowed ends.	. . . . . .
AURANTIACA.	Sarcina.	Small cocci in pairs and tetrads; strongly ærobic.	Orange-yellow pigment.
AURANTIACUS.	Bacillus.	Motile, short thick rods, often in long threads.	Orange-yellow pigment.
AURANTIACUS.	Micrococcus.	Oval cocci in pairs and singly; immotile.	Orange-yellow pigment in water, alcohol, and ether; insoluble.
AUREA.	Sarcina.	Cocci in packets.	Golden-colored pigment; soluble in alcohol.
AUREUS.	Bacillus.	Straight motile rods lying parallel.	Golden-yellow pigment.
BALTICUS.	Bacillus.	Short rod.	Phosphorescence.
BIENSTOCKII.	Bacillus.	See *Putrificus*, coli.	
BILLROTHII.	Micrococcus (ascococcus).	Groups of cocci surrounded with capsule; zooglœa ærobic.	. . . . . .
BRUNNEUS.	Bacillus.	Motile rods.	Brown pigment.
BUTYRIC-ACID FERMENTATION.	Bacillus.	Large, slender motile rods in pairs; spores; facul. anærobin.	Diastase.

BACTERIA.—CONTINUED.

Culture Characters.	Actions.	Habitat.	Discoverer.
Light-yellow colonies; serrated edges.	. . . . . .	Old distilled water.	Bolton.
. . . .	. . . . . .	Sulphur water.	Agardh.
Colonies, radiating from an oval centre like roots; later on colored yellow; slowly liquefy.	. . . . .	London Water-works.	Franeland.
. . . .	. . . . . .	Stagnant water.	Warming.
Rapidly liquefy; little orange-yellow colonies, not growing in high temperature.	. . . . . .	Air and water	Koch.
Slowly growing; nail cultures; shining and orange-yellow; not liquefy.	. . . . . .	Water.	Francland.
Round orange-yellow colonies, mostly on surface; slow growth; not liquefying.	. . . . . .	Water.	Cohn.
Liquefy; bright golden layer on potato.	. . . . . .	Exudate of pneumonia.	Mace.
Slow-growing, chrome-yellow, whetstone in shape; not liquefy.	. . . . . .	Water and skin of eczema.	Adametz and Unna.
Do not liquefy; require glucose for growth.	. . . . . .	Baltic Sea.	Fischer.
Creamy layer on surface of gelatin.	. . . . . .	Putrid broth.	Cohn.
. . . .	. . . . . .	Maize.	Schröter.
Liquefy rapidily; gray veil on surface of potato.	Casein ppt. and changed into butyric acid; ammonia set free.	Air.	Hueppe.

NON-PATHOGENIC

Name.	Genus.	Biology.	Product.
BUTYRICUM (amylobacter).	Clostridium.	Thick motile rods enlarging for the spores; obligat. ærobic.	Amyloid substance.
CÆRULEUS.	Bacillus.	Rods in long chains.	Blue pigment, not soluble in water, alcohol, or acid.
CANDICANS (candidus).	Micrococcus.	Masses of cocci.	. . . . .
CAROTARUM.	Bacillus.	Threads of rods that bend in various directions; oval spores.	. . . . .
CATENULA.	Bacillus.	Motile rods with spores.	. . . . . .
CAUCASICUS.	Bacillus.	Motile rods, with spores in each end.	. . . . . .
CERASINUS SICCUS.	Micrococcus.	Very small cocci, singly and in pairs; ærob.	Cherry-red pigment.
CEREUS ALBUS.	Micrococcus.	Cocci in short chains and bunches, colored by Gram.	. . . . . .
CEREUS FLAVUS.	Micrococcus.	Straphylo. and strepto., and in zoogloea, colored by Gram.	. . . . . .
CHLORINUS.	Bacillus.	Large rods, motile, green-colored, due to chlorophyll; ærobic.	Green pigment, soluble in alcohol.
CHLORINUS.	Micrococcus.	Cocci in zooglœa.	Green pigment, soluble in alcohol and water.
CINNABAREUS.	Micrococcus.	Large oval cocci in pairs; ærobic.	Brown-red pigment; foul odor.
CITREUS.	Bacillus (asco.).	Straight and bent rods in bundles; motile.	Citron yellow pigment.
CITREUS.	Micrococcus.	Large round cocci in chains of eight and more.	Cream-colored pigment.
CITREUS CONGLOMERATUS.	Micrococcus.	Diplococci and tetrads; ærobic.	. . . . . .
CLAVIFORMIS.	Bacillus (Tyrothrix).	Small rods; spores; true anærobin.	. . . . . .

BACTERIA.—Continued.

Culture Characters.	Actions.	Habitat.	Discoverer.
Not cultivated.	Forms butyric acid in presence of lactic acid.	Air, earth, and water.	Prazmowski and Van Tiegham.
Liquefy; a deep-blue layer on potato.	. . . . . .	Water.	Smith.
Not liquefy; nail-shaped in test-tube.	. . . . . .	Air around old cultures.	Flügge.
Rapidly liquefy on surface, a network centre on potato; round, light gray; grow rapidly.	. . . . . .	Cooked carrots and beets.	A. Koch.
. . . .	Causes albumin to ferment.	Old cheese.	Duclaux.
. . . .	Ferments milk, producing the kefyr drink.	Kefyr; grain.	Kern.
On potato; rapidly-forming cherry-red scum, not developed on gelatin.	. . . . . .	Water	List.
Not liquefy; small wax-like drops; thick gray layer on potato; growth rapid.	. . . . . .	Pus.	Passet.
Not liquefy; dark-yellow colonies; wax-like appearance.	. . . . . .	Pus.	Passet.
Liquefy; greenish-yellow colonies.	. . . . . .	Water.	Engelman.
Yellow-green layer on gelatin.	. . . . . .	Boiled eggs.	Cohn.
Not liquefy; slow growth; bright-red points.	. . . . . .	Air and water.	Flügge.
Slow growth; after two weeks small yellow points which take various shapes on potato; citron-yellow layer; growth more rapid.	. . . . . .	Skin in eczema.	Unna and Tommasoli.
Dirty cream-colored colonies, which are raised and moist.	. . . . . .	Water.	List.
Lemon-yellow colonies.	. . . . . .	Dust and blenorrhagic pus.	Bumm.
. . . .	Ferments milk, giving rise to alcohol.	Fermenting albumin.	Duclaux.

NON-PATHOGENIC

Name.	Genus.	Biology.	Product.
CONCENTRICUM.	Spirillum.	Thick motile spirals with flagella; ærobic.	. . . . . .
CORONATUS.	Micrococcus.	Cocci singly and streptococci; ærobic.	. . . . . .
CORYZÆ.	Micrococcus.	Large diplococci with rounded ends, the contact surfaces flat.	. . . . . .
CREPESCULUM. *	Micrococcus.	Round and oval cocci, singly and in zooglœa.	. . . . . .
CYANEUS.	Micrococcus.	Oval cells.	Blue pigment.
CYANOGENUS (blue milk).	Bacillus.	Motile rods in chains; spores; ærobic.	Alkali and a pigment deepened by acids
DICHOTOMA.	Cladothrix.	Various forms—rods, spirals, and cocci, in long threads.	. . . . . .
DIFFLUENS.	Micrococcus.	Oval cocci; ærobic.	Fluorescent pigment, soluble in water.
DISTORTUS.	Bacillus (Tyrothrix).	Motile rods; spores; ærobic.	Alkali.
DYSODES.	Bacillus.	Long and short rods; spores.	An odor resembling peppermint and turpentine.
ENDOPARAGOGICUM.	Spirillum.	Dry motile spirals, joined in peculiar shapes.	. . . . . .
ERYTHROSPORUS.	Bacillus.	Motile rods and threads; spores, slender.	Greenish - yellow pigment.
FIGURANS (mycoides).	Bacillus.	Large motile rods; spores; long threads; ærobic.	. . . . . .
FILIFORMIS.	Bacillus (Tyrothrix).	Short motile rods; spores in one end.	. . . . . .
FISCHERI.	Bacillus.	. . . . . .	Phosphorescence.
FITZIANUS.	Bacillus.	Short rods in threads; spores as large as the rods.	. . . . . .

BACTERIA.—Continued.

Culture Characters.	Actions.	Habitat.	Discoverer.
Not liquefying; concentrically-disposed colonies; very slow growth; not growing on potato.	. . . . . .	Putrefying blood.	Kitasato.
A halo formed around the colonies.	. . . . . .	Air.	Flügge.
White, raised glassy colonies, at first like pneumococci, later culture flattened; not liquefying.	No pathogenic action.	Acute coryzal secretion.	Hajek.
. . . .	. . . . . .	Putrefying infusions.	Cohn.
Bluish-green colonies.	. . . . . .	Cooked potatoes.	Cohn.
Not liquefying; small white colonies.	Changes milk to deep-blue color.	Air of certain countries.	Fuchs.
Cultivated in infusion of plants.	. . . . . .	Water.	Cohn.
Do not liquefy; small granular, yellow, colonies; green fluorescence.	. . . . . .	Air.	Schröter.
. . . .	Milk made viscid and casein precipitated.	Air.	Duclaux.
. . . . .	. . . . . .	Bread and yeast.	Zopf.
. . . .	. . . . . .	Trunk of worm-eaten tree.	Sorokin.
Does not liquefy; green fluorescence; white colonies.	. . . . . .	Air and putrefying substances.	Cohn.
Liquefying; root-like processes extending in the gelatin; feather form in test-tube.	. . . . . .	Garden-earth.	Flügge.
. . . .	Causes casein to be precipitated from milk.	. . . . . .	Duclaux.
Not liquefying; requires peptone for growth.	. . . . . .	. . . . . .	Beyerinck.
Transparent on surface; dark centre in the deep; not liquefying.	Produces ethylic alcohol in meat extract.	Unboiled hay-infusion.	Zopf.

NON-PATHOGENIC

Name.	Genus.	Biology.	Product.
FLAVA.	Sarcina.	Small cocci in packets.	Pigment.
FLAVUS.	Bacillus.	Small rods; immotile.	Pigment.
FLAVUS DESIDENS.	Streptococcus.	Cocci and diplococci in chains; ærobic.	Yellow-brown pigment.
FLAVUS LIQUEFACIENS.	Micrococcus.	Cocci and diplococci in zooglœa.	Pigment.
FLAVUS TARDIGRADUS.	Micrococcus.	Cocci in short chains, and diplococci.	Chrome-yellow pigment.
FLUORESCENS FŒTIDUS.	Micrococcus.	Small diplococci.	Blue-green pigment: acids turn red.
FLUORESCENS LIQUEFACIENS.	Bacillus.	Short motile rods; very thin.	Green fluorescent pigment.
FLUORESCENS NIVALIS.	Bacillus.	Short rods; motile.	Blue-green pigment.
FLUORESCENS PUTRIDUS.	Bacillus.	Motile rods; short, with rounded ends.	Green fluorescent pigment.
FOERSTERI.	Cladothrix.	Threads twisted in spirals; very irregular.	. . . . . .
FŒTIDUM.	Clostridium.	Rods of varying length; very motile; a large spore in one end; anærobic.	Strong gas-production; very foul odor
FŒTIDUS.	Micrococcus.	See *Crepesculum*, with	which it is identi
FUESCENS.	Sarcina.		
FULVUS.	Micrococcus.	Round cocci.	. . . . . .
FUSCUS LIMBATUS.	Bacillus.	Short rods; very motile; facultatively anærobic.	Brown pigment.
FUSIFORME.	Bacillus.	Spindle-shaped, with pointed ends.	. . . . .
GENICULATUS.	Bacillus (Tyrothrix).	Rods variable length; spores.	A bitter substance.
GIGANTEUS URETHRÆ.	Micrococcus.	Streptococci in thick knots.	. . . . . .

BACTERIA.—Continued.

Culture Characters.	Actions.	Habitat.	Discoverer.
Liquefying.	. . . . . .	Vomited matter.	
Liquefying; yellow viscid colonies; foul odor.	. . . . . .	Drinking-water.	Mace.
Yellow porcelain-white colonies.	. . . . . .	Air and old cultures; water.	Flügge.
Liquefying rapidly; yellow colonies.	. . . . . .	Air and old cultures; water.	Flügge.
Softens gelatin; yellow beads, isolated.	. . . . . .	Air.	Flügge.
Little button-like colonies that later on sink in, surrounded by violet-green color; liquefying; growth rapid.	. . . . . .	Post-nasal space.	Klamann.
Liquefying; white, sunken, iridescent colonies.	. . . . . .	Water and air; conjunctival sac.	Flügge.
Quickly liquefying; growth rapid; small white points; later on, surrounded by blue-green fluorescence.	Colors the glacial waters green.	In snow and ice of Norway.	Schmolck.
Not liquefying; transparent at first, then green fluorescence and urinary odor.	. . . . . .	All putrefactions.	Flügge.
. . . .	. . . . . .	Lachrymal canal.	Cohn.
Liquefying; growth rapid; small colonies that soon become filled up with fluid and assume a spherical form.	. . . . . .	Old cheese and serum of mice inoculated with garden-earth.	Liborius.
cal.			
Conical rusty-red colonies.	. . . . . .	Excrement of horse.	Cohn.
Small brown colonies, along needle-track little branches; not liquefy.	. . . . . .	In foul eggs.	Scheibenzuber.
. . . .	. . . . . .	Spongy layer on sea-water.	Warning.
. . . .	. . . . . .	Air and milk.	Duclaux.
No growth on gelatin; on agar, thin drops; nearly transparent; very slow growth; in bouillon, a flaky precipitate.	. . . . . .	Normal urine and urethra.	Lustgarten.

NON-PATHOGENIC

Name.	Genus.	Biology.	Product.
GRAVEOLENS.	Bacillus.	Small rods, nearly as broad as they are long.	Foul gas.
HÆMATODES.	Micrococcus.	Cocci in little zooglœa.	Red pigment.
HANSENII.	Bacillus.	Medium large rods.	Yellow pigment; insoluble.
HYACINTH.	Bacillus.	Short rods in dumb-bell shapes.	. . . . . .
HYALINA.	Sarcina.	Round cocci in groups of 4 to 24.	. . . .
IANTHINUS.	Bacillus.	See *Bacillus violaceus.*	
INDICUS.	Bacillus.	Short motile rods; no spores; anærobin facul.	Scarlet pigment altered by heat.
INTESTINALIS.	Sarcina.	Very regular packets of cocci, eight in each.	. . . . . .
JEQUIRITY	Bacillus.	Medium-sized rods; spores.	Ferment called abrin.
KÜHNIANA.	Crenothrix.	Long threads, breaking up into cocci. They are ensheathed.	. . . . . .
LACTEUS FAVIFORMIS.	Micrococcus.	Diplococci; not decolorized by Gram.	. . . . . .
LACTIS ERYTHROGENES.	Bacillus.	Short immotile rods; round ends.	Yellow pigment and red pigment.
LEPTOMITIFORMIS.	Beggiatoa.	Filaments medium size.	. . . . . .
LEUCOMELÆNUM.	Spirillum.	Two or three spirals; dark granular contents; clear spaces between.	. . . .
LINEOLA.	Bacillus.	Short motile rods in zooglœa, with flagella.	. . . . . .
LIODERMOS.	Bacillus.	Short motile rods; rounded ends.	. . . . . .

BACTERIA.—Continued.

Culture Characters.	Actions.	Habitat.	Discoverer.
Liquefying; irregular grayish, later greenish, colonies, with very foul odor.	. . . . . .	Skin between toes.	Bordoni-Uffreduzzi.
Grows best on white of egg at 37° C.; red layer.	. . . . . .	Sweat of man.	Zopf.
On potato, a yellow growth which changes with age.	. . . . . .	Yellow skin of nutrient infusions.	Rasmussen.
. . . .	. . . . . .	Slime of diseased hyacinth-bulbs.	Wakker.
. . . .	. . . . . .	Marshes.	Kützing.
Liquefying; oval colonies; scarlet-colored.	. . . . . .	Intestine of monkey.	Koch.
. . . .	. . . . . .	Intestine of fowls.	Zopf.
. . . .	Ferment causes ophthalmia.	Infusion of jequirity bean.	Sattler.
Colonies brick-colored from oxide of iron.	. . . . . .	Drinking-water of wells.	Rabenhorst.
Not liquefying; white colonies; grow well on potato.	. . . . .	Mucus of vagina and uterus.	Bumm.
Small, round yellow dots, later on cup-shaped, with rose-colored periphery; liquefying.	. . . . . .	In red milk and fæces.	Hueppe and Grotenfeldt.
. . . .	. . . .	Sulphur waters.	Trévisan.
. . . .	. . . . . .	Water over rotting plants.	Perty.
Slimy layer on potatoes.	. . . . . .	Stagnant water.	Müller.
Liquefying, transparent, then thick layer on potato; like gum.	. . . . . .	Air and potatoes.	Flügge.

NON-PATHOGENIC

Name.	Genus.	Biology.	Product.
LITORALIS.	Merismopedia.	Cocci in groups of fours, containing sulphur.	. . . . . .
LITOREUS.	Bacillus.	Oval rods, never in chains or zooglœa.	. . . . . .
LIVIDUS.	Bacillus.	Medium-sized rods; motile.	Deep blue-black pigment.
LUTEA.	Sarcina.	Cocci singly and in fours.	Pigment citron-yellow.
LUTEUS.	Bacillus.	Short immotile rods, with large oval spores.	Pigment; soluble in water; acids intensify.
LUTEUS.	Micrococcus.	Oval cocci.	Pigment, not acted upon by acid or alkali.
LUTEUS.	Micrococcus.	Diplococci very motile.	Yellow pigment, turning brown-red.
MAIDIS.	Bacillus.	Rods with pointed ends; very motile; seldom in threads; oval spores.	. . . . .
MARSH.	Spirillum.	See *Plicatile*.	
MEGATERIUM.	Bacillus.	Large motile rods; spores; ærobic.	.
MELANOSPORUS.	Bacillus.	Rods; ærobic.	Black pigment, not acted upon by acids or alkalies.
MERISMO-PEDIOIDES.	Bacillus.	Threads of rods which are formed from cocci-like spores; zooglœa in packets.	. . . . .
MESENTERICUS FUSCUS (potato).	Bacillus.	Small motile rods with spores.	. . . . . .
MESENTERICUS VULGATUS (potato).	Bacillus.	Thick motile rods in threads; spores.	Diastase.

BACTERIA.—Continued.

Culture Characters.	Actions.	Habitat.	Discoverer.
. . . .	. . . . . .	Sea-water.	Oersted and Rabenhorst.
. . . .	. . . . . .	Sea-water.	Warming.
Ink-spot at first, slowly liquefying; blue-violet colored later on; slow growth.	. . . . . .	Berlin Water-works.	Plogge and Proskauer.
Not liquefying; little elevations; citron-yellow centre; yellow layer on potato.	. . . . . .	Air.	Schröter.
Not liquefying; irregular in form; golden-yellow colored.	. . . . . .	Air.	Flügge.
Do not liquefying; small citron-yellow colonies on potato.	. . . . . .	Air.	Schröter.
Round, light-yellow colonies, growing larger in a few days; on potato a slimy covering with mouldy odor; slowly liquefying.	. . . . . .	Water.	Adametz.
Gray points in deep, veil-like on surface; liquefying; on potato, a wrinkled skin of brownish color.	In solutions of sugar an aldehyde produced.	In maize and in pellegra; fæces.	Paltauf and Heider.
Yellow irregular masses; thick layer on potato.	. . . . . .	Cooked cabbage.	De Bary.
First gray, then black, pellicle.	. . . . . .	Air and potatoes.	Eidam.
. . . .	. . . . . .	Stagnant water.	Zopf.
Liquefying; white colonies, ray-like periphery; brown layer on potato.	. . . . . .	Potato.	Flügge.
Yellow colonies, dark centre, ciliary processes at periphery; brown layer on potato, penetrating the substance.	Coagulates milk and forms diastase out of starch.	Air and old potatoes.	Flügge.

NON-PATHOGENIC

Name.	Genus.	Biology.	Product.
MESENTEROIDES.	Leuconostocci.	Masses of cartilaginous zoogloea, composed of rods and cocci; arthrospores.	.
MILLER'S.	Bacillus.	Delicate rods, slightly curved; immotile.	. . . . . .
MINUTA.	Sarcina.	Cube-shaped packets.	. . . . . .
MIRABILIS.	Beggiatoa.	Very wide threads, rounded ends and curled; sulphur granules.	. . . . .
MULTIPEDICULOSUS.	Bacillus.	Long, slender rods.	. . . . .
MULTISEPTATA.	Phragmidiothrix.	Long threads, containing cocci which are not free; they have no sulphur, and are not enclosed in a sheath.	. . . . .
NASALIS.	Micrococcus.	Diplococci, motile; also streptococci.	. . . . . .
NAVICULA.	Bacillus.	Spindle-shaped rods.	Amyloid material.
NITRIFICANS.	Micrococcus.	Small cocci.	Forms saltpetre.
NIVEA.	Beggiatoa.	Very thin filaments.	. . . . . .
NODOCUS PARVUS.	Bacillus.	Rods formed at angles; immotile.	. . . . . .
OBLONGUS.	Micrococcus.	Motile cocci, singly and in filaments; ærobic.	. . . . . .
OCHROLEUCUS.	Micrococcus.	Cocci in pairs and packets; spores.	Yellow pigment.
PALUDOSA.	Sarcina.	Spherical, transparent, colorless cocci.	. . . . . .

BACTERIA.—CONTINUED.

Culture Characters.	Actions.	Habitat.	Discoverer.
. . . .	Converts molasses into a gelatinous mass.	Beet-root juice.	Cienkowski.
Liquefies; not growing on the surface.	. . . . . .	Caries of teeth.	Miller.
Grows slowly; reacts to iodine, turning blue.	. . . . . .	Sour milk.	De Bary.
. . . .	. . . . .	Sea-water.	Cohn.
Insect-shaped colonies.	. . . . . .	Potatoes.	Flügge
. . . .	. . . . . .	Sea-water.	
Grayish points, raised, opaque; rapid growth; not liquefying.	. . . . . .	Nasal space and secretion.	Hack.
. . . .	. . . . . .	Potatoes.	Reinke and Berthold.
. . . .	. . . . . .	Soil.	Van Tieghem.
White flakes.	. . . . . .	Sulphur waters.	Rabenhorst.
Slow growth at 37° C.; in agar a white line, which in the centre becomes porous.	. . . . . .	Urethral secretion.	Lustgarten.
Grows best in cultures to which glucose and ammon. tartrate have been added.	Causes gluconic fermentation.	Beer.	Boutroux.
Liquefying; slow growth; thin yellow membrane; sulphurous odor.	. . . . . .	Urine.	Prove.
. . . .	. . . . . .	Water from sugar-factory.	Schröter.

NON-PATHOGENIC

Name.	Genus.	Biology.	Product.
PASTEURIANUS.	Bacillus.	Differs from bacil. aceti in that the cells contain an amyloid matter.	. . . . . .
PFLÜGERI.	Bacillus.	Short rods in threads.	Phosphorescence.
PHOSPHORESCENS GELIDUS.	Bacillus.	Motile; round, short rods; ærobic.	Phosphorescence.
PHOSPHORESCENS INDICUS.	Bacillus.	Large motile rods.	Phosphorescence.
PHOSPHORESCENS, North Sea.	Bacillus.	Motile rods.	Phosphorescence.
PHOTOMETRICUS.	Bacillus.	Motile, red-colored rods.	Sulphur and red pigment caused by light.
PLICATILE.	Spirillum.	Long motile, thin spirals; round ends.	. . . . . .
POLYMYXA.	Clostridium.	Motile rods in threads with spores.	Amyloid, colored blue by iodine.
PRODIGIOSUS.	Bacillus.	Short motile rods; ærobic.	Red pigment, soluble in alcohol trimethylamine.
PROTEUS MIRABILIS.	Bacillus.	Very motile, short rods; ærobic.	. . . . . .
PROTEUS VULGARIS.	Bacillus.	Rods sometimes curved, as spirillum.	. . . . . .
PROTEUS ZENKERI.	Bacillus.	Motile rods.	. . . . . .
PSEUDO-DIPHTHERIÆ.	Bacillus.	Small rods, similar to the true bacillus; immotile.	. . . . . .
PUTRIFICUS COLI.	Bacillus.	Slender motile rods; long threads; spores.	. . . . . .
PYOGENES TENUIS.	Micrococcus.	. . . . . .	. . . . . .
RADIATUS.	Bacillus.	Motile rods with rounded ends; anærobic; oval spores.	Strong - smelling gas.
RADIATUS.	Streptococcus.	Small cocci in chains.	. . . . . .

BACTERIA.—CONTINUED.

Culture Characters.	Actions.	Habitat.	Discoverer.
. . . .	. . . . . .	Heavy beers.	Hansen.
Not liquef'g; requires glucose; grows well on potato.	. . . . . .	Putrid meat and fish.	Ludwig.
Not liquefying; grows best with glucose and salt.	. . . . . .	Salt fish.	Förster.
Liquefying; grows best at 30° C.	. . . . . .	Tropical seas.	Fischer.
Liquefying; colonies look as if punched out; grows best at 15° C.	. . . . . .	Water around Kiel.	Fischer.
Movements depend upon light.	. . . . . .	. . . . .	Engelman.
. . . .	. . . . . .	Stagnant water.	Ehrenberg.
Thick skin on potato.	Causes fermentation in dextrin solutions.	. . . . . .	Prazmowski.
Little red colonies; liquefying rapidly; especially abundant on potatoes.	. . . . . .	Bread and potatoes.	Ehrenberg.
Liquefying slowly; opaque centre, irregular processes.	. . . . . .	Putrefaction.	Hauser.
Liquefying quickly.	. . . . . .	Putrefaction.	Hauser.
Not liquefying; thick white layer on potato.	. . . . . .	Putrefaction.	Hauser.
Grows at ordinary temperature, rapidly forming on surface a brownish growth; pin-head colonies raised above surface; not liquefying.	Not virulent.	In diphtheritic membrane and normal pharynx.	Wellenhof.
. . . .	Decomposes albumen.	Human fæces.	Bienstock.
On agar, a glassy growth.	. . . . . .	Closed abscesses.	Rosenbach.
Liquefying; growth rapid; colonies like moulds, from centre radiating in all directions and through the gelatin; the air must be excluded.	Not pathogenic.	In serum of white mice inoculated with earth.	Lüderitz.
Liquefying; white colonies with greenish tinge; funnel-shaped in test-tube.	. . . . . .	Air.	Flügge.

NON-PATHOGENIC

Name.	Genus.	Biology.	Product.
RAMOSUS LIQUEFACIENS.	Bacillus.	Motile rods.	. . . . . .
REITENBACHII.	Merismopedia.	Cocci in packets or plates; colorless cell-wall containing chlorophyll.	. . . . . .
ROSACEUS.	Micrococcus.	Large cocci in pairs and tetrads.	Red pigment.
ROSEA.	Sarcina.	Spherical cocci in cubical packets.	. . . . .
ROSEA PERSEINA.	Beggiatoa	Long rods with cocci-shaped bodies in them, containing sulphur and a red pigment.	Pigment called bacterio-purpurin.
ROSEUM.	Spirillum.	Very short curved rods; motile and spores.	Pigment soluble in alcohol.
RUBER.	Bacillus.	Motile rods in groups.	Brick-red pigment.
RUBRUM.	Spirillum.	Motile; short spirilla; ærobic.	Pale-rose pigment.
RUFUM.	Spirillum.	Long motile spirals.	Red-rose pigment.
RUGULA.	Spirillum (vibrio).	Motile rods, in long spirals, singly and in chains, with flagella and spores; anærobic.	. . . . . .
SAPROGENES.	Bacillus.	Large rods, terminal spores; facultatively anærobic.	. . . . .
SCABER.	Bacillus (Tyrothrix).	Short motile rods in chains; spores; ærobic.	Tyrosin and leucin are formed.
SCHEURLEN'S.	Bacillus.	Short motile rods; spores.	. . . . . .
SEPTICUS.	Bacillus.	Non-motile rods in threads and spores; anærobic.	. . . . . .
SERPENS.	Spirillum.	Long, lively threads, with three windings.	. . . . . .

BACTERIA.—Continued.

Culture Characters.	Actions.	Habitat.	Discoverer.
Liquefying; concentrical colonies; funnel-shaped in test-tube.		Air.	Flügge.
. . . .	. . . . . .	. . . . . .	Caspary.
Not liquefying; small red knobs, with fæcal odor.	. . . . . .	Air.	Flügge.
. . . .	. . . . . .	Marshes.	Schröter.
. . . .	. . . . . .	Marshes.	Zopf.
Not liquefying; thick violet colonies; deep red on potato.	. . . . . .	Blennorrhagic pus.	Mace.
. . . .	. . . . . .	Boiled rice.	Frank.
Not liquefying; grows slowly; pale-rose colonies.	. . . . . .	Dead mice.	Esmarch.
. . . .	. . . . . .	Stagnant water.	Perty.
Liquefying rapidly; round yellow dots with zone; fæcal odor.	Causes cellulose to ferment.	Vegetable infusions and tartar of teeth.	Müller.
Grows slowly; foul odor.	. . . . . .	Putrefaction.	Rosenbach.
. . . .	. . . . . .	. . . . . .	Duclaux.
Growth best at 39° C.; slowly liquefying on potato; a yellow, wrinkled skin, underneath which a red color.	. . . . . .	In carcinomatous and normal mamma.	Scheurlen.
. . . .	. . . . . .	Putrid blood.	Klein.
. . . .	. . . . . .	Stagnant water.	Müller.

NON-PATHOGENIC

Name.	Genus.	Biology.	Product.
SIMILIS.	Bacillus.	Immotile rods; transparent spores.	. . . .
SPINOSUS.	Bacillus.	Large motile rods; spores; true anærobin.	. . . .
SUBFLAVUS.	Micrococcus.	Diplococci colored by Gram's fluid.	. . . .
SUBTILIFORMIS.	Bacillus.	Immotile rods in threads; transparent spores.	. . . .
SUBTILIS (hay bacillus).	Bacillus.	Large motile rods, three times longer than broad, in threads, with flagella and spores; ærobic.	. . . .
SYNCYANEUS.	Bacillus.	Same as *Cyanogenus*.	
SYNXANTHUS (yellow milk).	Bacillus.	Short, thin motile rods.	Yellow pigment, soluble in water; similar to aniline colors.
TENUE.	Spirillum.	Large motile spirals with flagella.	. . . .
TENUIS.	Bacillus (Tyrothrix).	Motile rods in long chains; spores.	. . . .
TERMO.	Bacillus.	Short motile; cocci-like rods in zoogiœa.	. . . . . .
TREMULUS.	Bacillus.	Motile rods with flagella and large round spores.	. . . .
TUMESCENS.	Bacillus.	Short rods with spores.	. . . .
TURGIDUS.	Bacillus (Tyrothrix).	Short immotile rods in long chains; spores; ærobic.	Carbonate of ammonium.
ULNA.	Bacillus.	Very large rods in chains and singly; not very motile; large spores.	. . . .
UNDULA.	Spirillum.	Long motile spirals, with flagella.	. . . .
UREÆ	Bacillus.	Short rods; spores; ærobic.	Ferment, propylamine.
URINÆ.	Sarcina.	Small cocci in families.	. . . .

BACTERIA.—CONTINUED.

Culture Characters.	Actions.	Habitat.	Discoverer.
Grows rapidly.	. . . . . .	Human fæces.	Bienstock.
Liquefying; spiny periphery; foul odor due to methylmercaptin.	Albuminous decomposition.	Garden-earth.	Lüderitz.
Liquefying; yellow dots.	. . . . . .	Vaginal secretion and lochial discharges.	Bumm.
Grows best at 37° C.	. . . . . .	Human fæces.	Bienstock.
Liquefying; gray centre, wreath-like border; thick layer on potato.	. . . . .	Soil and dust, hay, etc.	Ehrenberg.
In boiled milk a yellow pigment is formed.	. . . . . .	Boiled milk and potatoes.	Ehrenberg.
. . . .	. . . . . .	Stagnant water.	Ehrenberg.
. . . .	Precipitates casein; forms a pellicle on milk.	Fermenting cheese and milk.	Duclaux.
Liquefying; opaque centre, yellow layer next, and the periphery lobed; funnel-shaped in test-tube.	. . . . . .	Connected with putrefaction of plants.	Dujardin.
. . . .	. . . . . .	Putrefying plants.	
On boiled carrots a wrinkled gelatinous disk.	. . . . . .	Boiled carrots.	Zopf.
A pellicle formed on surface of milk; a heavy precipitate beneath.	. . . . . .	Fermenting milk and cheese.	Duclaux.
On boiled egg little zooglœa.	. . . . . .	Putrefying water and boiled eggs.	Cohn.
. . . .	. . . . . .	Vegetable infusions.	Müller.
Resembling a globule of fat; grows well in mucous urine.	Splits urea into ammonii carbonas.	Stale urine.	Miquel.
. . . .	. . . . . .	Bladder.	Welcker.

NON-PATHOGENIC

Name.	Genus.	Biology.	Product.
UROCEPHALUS.	Bacillus (Tyrothrix).	Cylindrical motile rods with spores; anærobic.	. . . . . .
VENTRICULA.	Sarcina.	Cubical packets of 8 to 64 cocci.	. . . . . .
VENTRICULI.	Bacillus.	Rods motile, often in bundles of four.	. . . . . .
VERSICOLOR.	Micrococcus.	Small cocci.	. . . . . .
VIOLACEUS.	Bacillus.	Motile rods, round end; spores.	Violet pigment, soluble in alcohol.
VIOLACEUS.	Bacillus.	Immotile rods, forming large spores.	Violet pigment, like aniline.
VIRENS.	Bacillus.	Straight rods; spores; immotile; green tinged.	Supposed to contain chlorophyll.
VIRESCENS.	Bacillus.	Short motile rods with flagella very broad.	Deep-green pigment, turning yellow-brown.
VIRGULA.	Bacillus (Tyrothrix).	Slender immotile rods; spores ærobic.	. . . . . .
VIRIDIS.	Bacillus.	Little immotile rods; oval spore, which is tinged green.	. . . . .
VISCOSUS.	Bacillus.	Motile rods, rounded ends, usually in pairs.	Green pigment.
VISCOSUS.	Micrococcus.	Streptococci of globular cells.	Gummy substance, called viscosa, and ferment.
VITICULOSOS.	Micrococcus.	Oval cocci in large groups.	. . . . . .
VOLUTANS.	Spirillum.	Long spirals with flagella.	. . . . . .
ZOPFI.	Bacillus.	Long motile rods, breaking up into spores like cocci.	. . . . . .

BACTERIA.—CONTINUED.

Culture Characters.	Actions.	Habitat.	Discoverer.
. . . .	. . . . . .	Fermenting milk.	Duclaux.
Not liquefying.	. . . . . .	Contents of stomach.	Goodsir.
Round colonies with dark centre; slow growth; not liquefying.	Peptonizes albumen.	Stomach of dogs fed on meat.	Raczynssky.
Not liquefying; iridescent yellow surface.	. . . . . .	Air.	Flügge.
Not liquefying; centre deep violet; color remains on agar a long time.	. . . . . .	Water.	Zopf.
Liquefying; transparent colonies, surrounded by violet zone.	. . . . . .	Boiled potato and water.	Schröter.
. . . .	. . . . . .	Stagnant water	VanTiegham.
Deep round colonies, the vicinity colored green; grows on surface; slow growth; not liquefying.	. . . . . .	Green sputum.	Frick.
. . . .	. . . . . .	Milk.	Duclaux.
. . . .	. . . . . .	Water.	VanTiegham.
Rapid growth, liquefying; small hair-like processes from colonies; later on, viscid and in threads, with green fluorescence.	. . . . .	Water and earth.	Francland.
. . . .	Mucoid fermentation in wine and beer.	Beer and wine.	Pasteur.
Not liquefying; a fine network in the colony; mucoid layer on potato.	. . . . . .	Air.	Flügge.
. . . .	. . . . . .	Marshes.	Ehrenberg.
Not liquefying; forms thick coils like braided hair.	. . . . . .	Intestinal contents of fowls.	Kurth.

PART II.

Name.	Genus.	Biology.	Product.
ACUTE YELLOW ATROPHY.	Micrococcus.	. . . .	. . . .
ALVEI.	Bacillus.	Rods with large spores.	. . . .
AMYLIVORUS.	Micrococcus.	Oval cells, never in chains.	Forms butyric acid
ANTHRAX.	Bacillus.	Straight rods, slightly concave ends; immotile; ærobic; spores.	Toxalbumin.
ARTICULORUM (diphtheriticus).	Micrococcus.	Oval cocci in long chains, identical with pyogenes.	. . . .
BISKRA BOIL (Aleppo boil).	Micrococcus.	Cocci united often in large numbers; immotile; capsules around diplococci.	. . . .
BOMBYCIS.	Micrococcus.	Oval cocci in chains and zooglœa; motile.	. . . .
BUCCALIS.	Leptothrix.	Long threads in thick bundles, containing masses of cocci and spirals.	. . . .
CATTLE PLAGUE (Texas fever).	. . . . . .	See *Hæmorrhagic Septi*	*cæmia* and *Swine*
CAVICIDA.	Bacillus.	Little rods twice as long as broad.	Propionic acid through decomposition of sugar.
CHAUVÆI (symptomatic anthrax), (Rauschbrand).	Bacillus.	Large rods with a spore at one end, assuming the clostridium type; motile; never in threads; true anærobin.	Toxalbumin.
CHOLERA ASIATICÆ	Spirillum.	Motile spiral-shaped rods, often in chains; very short flagella on ends, and strictly ærobic; spores have not been found.	Ptomaïne-like muscarine; and toxalbumin, soluble in water.

PATHOGENIC BACTERIA.

Culture Characters.	Actions.	Habitat.	Discoverer.
. . . .	. . . . . .	Liver of yellow atrophy.	Eppinger.
Liquefying; growths radiating from centre downward; on potato a dry yellow layer.	Produces a disease in bees called "foul brood."	Larvæ of bees.	Chesbire and Cheyne.
. . . .	"Fire-blight" in pear trees.	. . . . . .	Burrill.
Liquefying; granular colonies with irregular border; on potato a dry, creamy layer; in test-tube a thorny, prickly track.	Causes splenic fever in animals; malignant pustule in man.	Found in tissues and excreta of diseased animals.	Rayer and Davaine.
Grows well on gelatin; pale-gray colonies; not liquefying; slow growth on potato.	Fatal in mice and rabbits.	Mucous membrane of diphtheria.	Löffler and Cohn.
Liquefying; light-yellow colonies; grow quickly.	Produces the Aleppo or Biskra boil, common in Africa and Asia.	Blood of the disease.	Duclaux and Heydenreich.
. . . .	Causes "flacherie" in silkworms.	Intestines of silkworms.	Béchamp.
. . . .	Causes dental caries.	Teeth slime.	Robin.
Plague.			
Not liquefying; irregular scale-like colonies, making the gelatin viscid.	Kills guinea pigs.	Human fæces.	Brieger.
Liquefying; opaque centre with ragged periphery; in test-tube growth below, with gas formation.	Causes "black leg," or Rauschbrand, in cattle.	Animals affected with disease.	Arloing, Carnevin, and Thomas.
Liquefying slowly, small depressed scars giving a frosted appearance, or like ground glass; on potato, a thin brown layer; in test-tube, a funnel-shaped liquefaction, with a bubble of air in the top, the funnel taking six or seven days to form well.	Causes cholera Asiatica in man and a similar trouble in animals.	Fæces of cholera patients.	Koch.

PATHOGENIC

Name.	Genus.	Biology.	Product.
CHOLERA GALLINARUM (chicken cholera).	Bacillus.	Immotile, cocci-like rods; without spores; strictly ærobic.	Toxalbumin.
CHOLERA NOSTRAS (Finckler).	Spirillum.	Motile, comma-shaped rods; strictly ærobic.	. . . . . .
COLI COMMUNIS.	Bacillus.	Short motile rods, slightly curved, without spores; facultatively anærobic.	. . . . . .
CRASSUS SPUTIGENUS.	Bacillus.	Short, thick rods with rounded ends.	. . . . . .
DECALVENS.	Micrococcus.	Spherical cells in great numbers.	. . . . . .
DENTALIS VIRIDANS.	Bacillus.	Slightly curved rods, round ends.	Gray pigment.
DIARRHŒA OF INFANTS.	Bacillus.	Motile, medium-sized rods; spores; ærobic.	Toxalbumin.
DIARRHŒA OF MEAT-POISONING.	Bacillus.	Rods in groups of two and singly; round ends; spores.	. . . . . .
DIPHTHERIÆ.	Bacillus.	Immotile, middle-sized rods, rounded ends; facultat. anærobic.	Toxalbumin.
DIPHTHERIA OF CALVES (Vitulorum).	Bacillus.	Long rods in threads.	. . . . . .
DIPHTHERIA IN PIGEONS (Columbarum).	Bacillus.	Short rods in groups.	. . . . . .

BACTERIA.—CONTINUED.

Culture Characters.	Actions.	Habitat.	Discoverer.
Not liquefying; small isolated white disks; in test-tube, a granular track; very faint.	Causes chicken cholera in fowls; not acting on man.	Blood and fæces of diseased fowls.	Pasteur.
Liquefying rapidly; colonies yellow-brown thick masses; in test-tube, funnel formed in 24 hours, dissolving all gelatin in two days; profuse gray mass on potato.	Harmless in man; fatal to guinea pigs.	Fæces of cholera nostras and caries of teeth.	Finckler and Prior.
Not liquefying; dark centre, undulated periphery; green-colored layer on potato; milky layer on surface of test-tube.	Fatal to guinea pigs and rabbits; causes diarrhœa in man.	Fæces of nursing infants; water; choleraic stools.	Escherich.
Not liquefying; oval grayish, slimy colonies; nail-shaped growth in test-tube.	Mice and rabbits die in 48 hours with gastro-enteritis.	Sputum.	Kreibohm.
. . . . . .	Causes alopecia areata.	In roots of hair.	Thin.
Not liquefying; round, sharply-outlined colonies, with bluish-gray opalescence.	Septic processes and death in mice and pigs.	In caries of teeth.	Miller.
Not liquefying; green colonies with foul odor.	Causes green diarrhœa in animals when intravenously injected, and is the cause of green diarrhœa in infants.	Fæces of infants suffering from green diarrhœa.	Lesage.
. . . .	Causes death in animals, with symptoms of septicæmia.	Blood and juices of choleraic diarrhœa.	Klein.
Not liquefying; little yellowish colonies; a membranous layer on potato.	Gives rise to diphtheria in man and animals.	Diphtheritic exudate.	Löffler.
. .	When inoculated in mice causes death.	Diphtheritic membrane of calf.	Löffler.
Whitish patches.	Necrosis in pigeons and other animals.	Diphtheritic membrane in pigeons.	Löffler.

PATHOGENIC

Name.	Genus.	Biology.	Product.
Duck Cholera.	Bacillus.	Similar to chicken cholera bacillus; immotile.	......
Dysentery (epidemic).	Bacillus.	Short motile rods; very thin.	......
Erysipelas of Swine (Rothlauf; rouget du porc).	Bacillus.	Small, slender motile rods; facultatively anærobic.	Two vaccines, which give immunity.
Fœtidus ozænæ.	Bacillus.	Short rods, very motile; in pairs and chains.	Foul gas.
Frog Plague.	Bacillus.	See *Swine Plague*.	......
Gangrene.	Micrococcus.	Oval cocci in zooglœa.	......
Gigantea.	Leptothrix.	Long rods, cocci and short rods in one; thread also spiral.	......
Gingivæ pyogenes.	Bacillus.	Short thick rods with rounded ends.	......
Glanders (Rotz, Mallei).	Bacillus.	Slender, immotile rods, usually singly; spores; facultatively anærobic.	......
Gonorrhœa (Gonococcus).	Micrococcus.	Diplococci kidney-shaped; motile; do not color with Gram.	......
Grouse Disease.	Bacillus.	Small rods and oval cocci in chains; immotile.	......
Hæmatococcus bovis.	Diplococcus.	Cocci seldom in chains; surrounded by a pale zone.	......
Hæmophilia neonatorum.	Micrococcus.	......	

BACTERIA.—CONTINUED.

Culture Characters.	Actions.	Habitat.	Discoverer.
Small round yellow colonies like wax-drops; not liquefying.	Fatal for ducks, but not for chickens or pigeons; less active than chicken cholera; causes diarrhœa and exhaustion.	Blood of diseased ducks.	Cornil and Toupet.
Not liquefying; concentrically-arranged colonies; dry yellow membrane on potato.	The cause of epidemic dysentery in man; enteritis in guinea pigs.	In fæces and mesenteric glands.	Chantemesse and Widal.
Very delicate silver-gray clouds on the gelatin, like bone-cells; not liquefying; in test-tube a very faint clouding.	Causes erysipelas in swine and other animals; the German "Rothlauf," French "rouget du porc."	Blood and organs of diseased animals.	Löffler.
Small greenish colonies which soon become liquefied and indistinguishable; a foul odor produced.	Mice are killed by injection; rabbits affected with progressive gangrene.	Secretion of persons suffering from ozæna.	Hajek.
. . . .	. . . .	. . . . . .	Eberth.
Grayish colonies with foul odor.	. . . . . .	Gangrenous tissue.	
. . . .	Causes caries of teeth.	Diseased teeth of animals.	Miller.
Growth rapid; liquefying; round colonies, visible to naked eye in 24 hours.	Fatal to mice, with septic processes.	Suppurating pulp of tooth.	Miller.
Light yellow, like honey, colonies, turning red-brown in a few days.	Glanders is caused by the bacillus in man and animals.	In epithelium and ulcerated glands.	Löffler.
Grow on blood-serum.	Gonorrhœa in man.	Gonorrhœal pus; in pus-cells and epithelium.	Neisser.
Not liquefying; small scales which turn gray in a few days, the edges serrated.	Fatal for mice and guinea pigs.	In blood and organs of diseased grouse.	Klein.
Best at 38° C.; not liquefying; small white points; sparse growth on potato; transparent.	Fatal for rabbits and rats; hyperæmia of lungs and spleen; blood - exudate in peritoneal cavity.	Blood and organs of animals diseased with hæmoglobinuria.	Babes.
. . . .	Supposed to be the cause of the disease.	Found in this disease.	Klebs.

PATHOGENIC

Name.	Genus.	Biology.	Product.
HÆMORRHAGIC SEPTICÆMIA (Infectious Pleuro-pneumonia, Wild Plague, German Swine Plague, Cattle Plague, Steer Plague, Rabbit Septicæmia).	Bacillus.	Short rods, twice as long as broad; immotile.	. . . .
HOG CHOLERA (Swedish swine plague).	Bacillus.	Very motile oval rods, similar to hæmorrhagic septicæmia.	Peptonizes milk without coagulation.
INSECTORUM.	Micrococcus.	Oval cells in chains and zooglœa; streptococci.	. . . . . .
LACTIS ÆROGENES.	Bacillus.	Short, thick immotile rods.	. . . . . .
LEPRÆ.	Bacillus.	Slender, immotile rods with pointed ends.	. . . . . .
LIQUEFACIENS CONJUNCTIVÆ.	Micrococcus.	Single cocci; never in threads.	. . . . . .
LUPUS.	Bacillus.	Same as *Tuberculosis.*	
MALARIA.	Bacillus.	Rods in filaments, with spores at each end; ærobic.	. . . . . .
MALIGNANT ŒDEMA (Gangrenous Septicæmia, Vibrio Septique).	Bacillus.	Large, slender rods, rounded ends, often in threads; motile, with flagella and spores; strongly anærobic.	Soluble vaccine.
MAMMITIS OF COWS.	Micrococcus.	Oval cocci in chains; streptococci; facultatively anærobic.	. . . . . .

BACTERIA.—CONTINUED.

Culture Characters.	Actions.	Habitat.	Discoverer.
White isolated pinhead points, not growing on potato; best at 37° C.; not liquefying.	A disease having different names in different animals, characterized by œdema, hæmorrhage, and septicæmia.	Blood and serum of diseased animals.	Hueppe.
Very good growth on gelatin and potatoes; a yellow-brown color.	In experiment, animal's death in four to eight days; bacteria in little emboli in capillaries.	Not spread through tissue, but in capillaries of diseased swine.	Salmon and Selander.
. . . .	A contagious disease in the chinch-bug.	Stomach of chinch-bug.	Burrill.
Small porcelain-like disks with depressed centre; funnel-shaped in test-tube with gas.	Fatal to guinea pigs and rabbits; coagulates milk; decomposes sugary solutions.	Fæces of nursing infants and of cholerine.	Escherich.
On blood-serum round white plaques with irregular borders.	Causes leprosy in man and animals.	Leprous tissue.	Hansen.
Liquefying; growth rapid; colonies on surface, with little radiating branches from a dark centre; those in deep, berry-shaped.	On cornea of rabbits causes slight clouding.	Normal human conjunctiva.	Gombert.
. . . .	Produces fever in man and animals.	Blood of malaria and air of malarial districts.	Klebs and Tommasi-Crudeli.
Liquefying; thick centre, radiating periphery; in high culture in test-tube, gas-bubbles arise, with foul odor.	Animals quickly die with extensive gangrene and œdema.	Garden-earth.	Pasteur.
Not liquefying; brown, round granular colonies; grows slowly; in test-tube, heavy deposit along the needle's track.	Causes contagious mammitis in cows; coagulates milk.	Mammary gland.	Nocard and Mollereau.

PATHOGENIC

Name.	Genus.	Biology.	Product.
MAMMITIS OF SHEEP.	Micrococcus.	Streptococci and in fours.	. . . . . .
METSCHNIKOWI.	Spirillum (vibrio).	Motile spirals with flagella; ærobic.	An alkaline vaccine which will cause immunity.
MORBILLI.	Micrococcus.	Round motile cocci and diplococci.	. . . . .
NEAPOLITANUS.	Bacillus.	Small immotile rods, with rounded ends; no spores; facultatively anærobic.	Produces acids in gelatin cultures.
NOMÆ.	Bacillus.	Small rods, with rounded ends, growing often in long threads.	. . . . .
OXYTOCUS PERNICIOSUS.	Bacillus.	Short rods with round ends.	. . . . . .
PERTUSSIS.	Bacillus.	Very thin rods; motile; spores.	. . . .
PNEUMONIA (Pneumococcus of Friedländer).	Bacillus.	Short, immotile rods, singly or in diplococci, surrounded with capsule; no spores; not colored with Gram; facultatively anærobic.	. . . . . .
PNEUMONIA (Pneumococcus of Fränkel; Micrococcus of Pasteur).	Bacillus.	Short, oval rods, often in chains; immotile; no spores; in the tissue surrounded with capsule, colored with Gram; facultatively anærobic.	. . . . .
PNEUMONICUS AGILIS.	Bacillus.	Short, thick motile rods in pairs.	. . . . . .
PROTEUS SEPTICUS.	Bacillus.	Slightly curved rods, swelled in portions, sometimes in long threads; motile.	Foul gas.

BACTERIA.—CONTINUED.

Culture Characters.	Actions.	Habitat.	Discoverer.
Liquefying; round centres with zone of liquefaction; cone-shaped in test-tube.	Causes contagious gangrenous mammitis in sheep.	Found in the milk of diseased sheep.	Nocard.
Grows quickly; colonies, some like cholera Asiatica, others like cholera nostras; liquefying.	Causes vibrion septicæmia in guinea pigs and pigeons.	Fæces of fowls.	Gamaleia.
. . . .	Supposed to have an intimate connection with measles.	Urine, blood, and catarrhal exudations of measles.	Keating.
Not liquefying; thin pearl-like scales in several layers; wrinkled and mucous layers on potato.	Causes death in some animals; not the cause of cholera.	Cholera epidemic of Naples, 1884.	Emmerich.
Granular spherical colonies in the deep, flat on the surface; not liquefying; growth rapid; best at 35° C.	No action on mice or rabbits.	In necrotic tissue of noma.	Schimmelbusch.
Small yellow granular colonies; nail-culture in test-tube.	Intravenous injection causes death in mice and rabbits; turns milk acid.	Sour milk.	Wyssokowitsch.
Not liquefying; thick yellow culture on potato.	Said to be constantly present in whooping cough; injected into trachea of young dogs, it produces broncho-pneumonia.	Phlegm of whooping cough.	Afanassieff.
Does not liquefy; grows quickly; a button-like colony; in test-tube, as if a nail driven in the gelatin with head on surface.	An accompaniment of pneumonia, not a cause; animals not affected.	Pneumonic and other sputum, and lung tissue.	Friedländer.
Does not liquefy; grows slowly; small, well-defined masses; in test-tube, little separate globules, one above the other.	Causes pneumonia in man, septicæmia in animals; also serous inflammations in man, as pleurisy, peritonitis, etc.	Sputum of lung affections and serous inflammations.	A. Fränkel.
Liquefying; dark granular colonies; thick sediment in test-tube.	Pneumonia in rabbits.	From rabbits' pneumonia.	Schon.
Growth rapid; liquefying; colonies have foul odor, are small, thick branches, but soon all liquid.	Fatal for mice in one to three days.	From a child dying of intestinal gangrene.	Babes.

, PATHOGENIC

Name.	Genus.	Biology.	Product.
PSEUDO-PNEUMONIA.	Bacillus.	Immotile, very short rods with capsule.	. . . . . .
PSITTACI (perniciosus).	Micrococcus.	Streptococci and zooglœa.	. . . . . .
PYOCYANUS.	Bacillus.	Thin motile rods; facultatively anærobic.	Pyocyanin, a non-poisonous pigment.
PYOCYANEUS β.	Bacillus.	Forms a brown-yellow pigment; otherwise identical with above.	. . . . . .
PYOGENES (Streptococcus erysipelatis—Fehleisen).	Micrococcus.	Streptococci and zooglœa.	. . . . . .
PYOGENES ALBUS.	Micrococcus.	Staphylococci and streptococci; facultatively anærobic.	. . . . . .
PYOGENES AUREUS (micrococcus of osteomyelitis—Becker).	Micrococcus.	Staphylococci and zooglœa; facultatively anærobic.	Ptomaïne, toxalbumin, and pigment.
PYOGENES CITREUS.	Micrococcus.	Same as *Pyogenes aureus.*	. . . . . .
PYOGENES FŒTIDUS.	Bacillus.	Short motile rods in pairs.	. . . . . .
PYOGENES TENUIS.	Micrococcus.	Cocci without definite arrangement.	. . . . . .
RABIES (Hydrophobia).	Bacillus.	Very thin rods.	Ptomaïne, which gives immunity when inoculated, and cures.
RELAPSING FEVER (Obermeier).	Spirillum.	Long, wavy spirals; motile.	. . . . . .

BACTERIA.—Continued.

Culture Characters.	Actions.	Habitat.	Discoverer.
Not liquefying; thick glistening brownish layer on potatoes.	Septicæmia in mice; abscess in guinea pigs.	From pus.	Passet.
. . . . . .	Causes disease in gray parrots.	In blood of parrot's disease.	Wolff.
Liquefying; large, flat colonies with greenish fluorescence; on potato, yellow-green skin, deeply coloring the pulp.	Fatal for animals; colors the dressings green.	Pus.	Gessard.
. . . .	. . . . . .	. . . . . .	Ernst.
Not liquefying; round punctiform colonies; slow-growing.	Suppuration and septicæmia in animals.	Pus.	Rosenbach.
Liquefying; white opaque colonies.	Suppuration and abscess.	Pus.	Rosenbach.
Liquefying; small colonies with a yellow-orange pigment in centre; yeast-like smell; a moist layer on potato.	Causes abscesses and suppuration in man and animals.	Pus.	Rosenbach.
Colonies, citron-yellow color.	Suppuration.	Pus.	Passet.
Not liquefying; mucous layer on potato; very thick; in test-tube, a slight layer on surface, and small points along the track.	Fatal to animals.	Pus.	Passet.
On surface, transparent; thin growth; grows slowly.	. . . . . .	Pus of abscesses.	Rosenbach.
A clouding in bouillon, which deposits itself in a few weeks.	Causes hydrophobia in animals.	From serum of ventricles and spinal cord.	Gibier, Mottet, and Protopnoff.
Cannot be cultivated.	Causes fever in man and animals, and is the cause of relapsing fever.	Blood of man during an attack of the disease.	Obermeier.

PATHOGENIC

Name.	Genus.	Biology.	Product.
RHINOSCLERMA.	Bacillus.	See *Pneumococcus* of	Friedländer, with
SALIVARUS PYOGENES.	Micrococcus.	Very small round cocci and staphylococci.	. . . . . .
SALIVARUS SEPTICUS.	Bacillus.	Short, immotile rods, encapsulated in pairs, sometimes long chain; ærobic.	. . . . . .
SALIVARUS SEPTICUS.	Micrococcus.	Cocci singly and in zooglœa; ærobic.	. . . . .
SAPROGENES No. II.	Bacillus.	Short rods; facultatively anærobic.	Foul gas.
SAPROGENES No. III.	Bacillus.	Very short rods; facultatively anærobic.	Foul gas.
SAPROGENES FŒTIDUS.	Bacillus.	Immotile rods; spores.	Foul gas.
SENILE GANGRENE.	Bacillus.	Thin rods; immotile; singly and in pairs; ends somewhat thickened; ærobic; spores.	. . . . . .
SEPTICÆMIA AFTER ANTHRAX.	Micrococcus.	Motile streptococci.	. . . . . .
SEPTICÆMIA OF MICE.	Bacillus.	Smallest bacillus known; immotile.	. . . . . .
SEPTICÆMIA OF RABBITS (Cuniculicida).	Bacillus.	See *Hæmorrhagic Septi*	*cæmia.*
SEPTICUS ACUMINATUS.	Bacillus.	Thin, lancet-shaped rods; very slender.	. . . . . .
SEPTICUS AGRIGENUS.	Bacillus.	Very short rods.	. . . . . .

BACTERIA.—CONTINUED.

Culture Characters.	Actions.	Habitat.	Discoverer.
which it is identical.	. . . . . .	. . . . . .	Frisch.
Slowly liquefying; small white opalescent colonies.	Local abscess in animals.	Saliva.	Biondi.
Not liquefying; gray circular colonies; transparent zone; in test-tube, separated.	Fatal to animals.	Saliva of healthy persons.	Biondi.
Not liquefying; round colonies; separated dots in test-tube.	Fatal to animals.	Saliva of puerperal women.	Biondi.
Grows quickly; on agar, hyaline drops which quickly coalesce, and form a mucoid layer with a foul odor, that of perspiring feet.	Produces septicæmia in rabbits.	Sweat of feet.	Rosenbach.
Forms a fluid gray band on agar; odor of putrefaction.	Suppuration in rabbit.	Putrid marrow of bone.	Rosenbach.
Not liquefying; thin, transparent layer; putrid odor.	Rabbits killed with large doses.	Mesenteric glands of swine with erysipelas and of healthy swine.	Schottelius.
Round yellow colonies; liquefying in 36 hours; best growth at 37° C.	Causes gangrene in mice, similar to senile gangrene of man.	In gangrenous tissue and blood of senile gangrene.	Tricomi.
In bouillon virulence destroyed.	Septicæmia in rabbits, but not in chickens or guinea pigs.	Blood of animal dead from anthrax.	Charrin.
Not liquefying; small flocculent masses in the deep; grows very slowly; in the test-tube producing a faint cloud.	Septicæmia in house-mice, but not field-mice.	Putrefying liquids.	Koch.
At 37° C. on blood-serum small transparent plates; later on, turning yellow.	Pathogenic for rabbits and guinea pigs; fever; and bacilli in blood and organs.	Navel stump of child dead of septicæmia.	Babes.
Not liquefying; brown centre, a ring, then yellow zone.	Septicæmia in mice and rabbits.	Earth of recently-ploughed fields.	Nicolaier.

PATHOGENIC

Name.	Genus.	Biology.	Product.
SEPTICUS LIQUEFACIENS.	Micrococcus.	Streptococci and diplococci.	. . . .
SEPTICUS ULCERIS.	Bacillus.	Oval rods; motile.	Gas; no odor.
SEPTICUS VESICÆ.	Bacillus.	Rods always single; very motile; oval spores.	. . . .
SMEGMA.	Bacillus.	Slender curved rods, identical with syphilis.	. . . .
SPUTIGENUM.	Spirillum.	Curved, comma-shaped rods; motile.	. . . .
SUBFLAVUS.	Micrococcus.	Diplococci like gonococci; colored by Gram.	. . . .
SWINE PLAGUE (American and French).	Bacillus.	Motile, oval rods, similar to that of hog cholera.	Causes casein precipitate in milk and acid formation.
SYCOSIFERUS FŒTIDUS.	Bacillus.	Short, straight immotile rods, often in threads.	On potatoes a foul odor.

BACTERIA.—CONTINUED.

Culture Characters.	Actions.	Habitat.	Discoverer.
Liquefying; a thin granular streak, the surface sunken in; later, cone-like, the walls covered with leaf-shaped colonies.	Pathogenic for mice and rabbits, producing œdema, in the serum of which the cocci abound.	Blood and organs of child dying of septicæmia.	Babes.
Liquefying; yellow colonies, taken up with gas later on.	An ulcer in inoculated animals, followed by paralysis and death.	In blood of child with gangrenous ulcer.	Babes.
Not liquefying; small pin-head colonies, growing slowly; never larger; a brown centre, yellow periphery.	Pathogenic for mice and rabbits, producing death.	In urine of cystitis.	Clado.
Not cultivated.	Supposed to be similar to syphilis.	Normal preputial secretions.	Alvarez and Tavel.
Not cultivated.	Causes death in animals.	In caries of teeth and saliva.	Lewis.
Growth slow; liquefying; on tenth day yellow points with thready boundary; on potato, a brown, thread-like growth after two weeks.	No result on mucous membrane; injected under skin, abscess results.	Normal secretion of vagina and urethra.	Bumm.
Not liquefying; growth similar to typhoid germ; on potatoes good growth.	Found in American and French swine plague, in frog plague, and Texas fever; animals affected locally.	Found in capillaries in little emboli; not spread in organs of diseased animals.	Billings, Rietsch, and Eberth.
Slow growth; not liquefying; after four days, little white points, which do not change for several weeks, then the superficial ones are mucous-like; nail growth; on potatoes, rapid growth.	On human skin causes eruption, vesicular around hairs, then it becomes pustular; similar to sycosis.	From sycosis of the beard.	Tommasoli.

PATHOGENIC

Name.	Genus.	Biology.	Product.
SYPHILIS.	Bacillus.	Thin rods, sometimes curved.	. . . . . .
TETANUS.	Bacillus.	Large, slender motile rods, with spores in one end, drumstick shape, often in threads; true anærobic.	Ptomaines, tetanine, tetanotoxine, spasmotoxine; also a toxalbumin.
TETRAGENUS.	Micrococcus.	Large round cells, united in groups, usually of four, and surrounded by a capsule; immotile; ærobic.	. . . . . .
TOXICATUS.	Micrococcus.	Cocci singly and in pairs.	. . . . . .
TRACHOMA	Micrococcus.	Diplococci very small and division - line faint; ærobic.	. . . . . .
TUBERCULOSIS.	Bacillus.	Slender rods, usually in pairs; not motile; spores not definitely determined; facultatively anærobic.	Kochine or paratoline, a glycerin extract of the pure culture (tuberculin).
TUBERCULOSIS ZOOGLŒAIC.	Micrococcus.	Cocci in large zoogleaic masses, evoluted forms of tubercle bacillus.	. . . . . .
TYPHOID.	Bacillus.	Slender motile rods, sometimes in threads; flagella, but no spores; facultatively anærobic.	Typhotoxin and toxalbumin.
TYPHOID OF SWINE (swine plague).	. . . . . .	See *Swine Plague*.	
TYROGENUM (Dencke's).	Spirillum (vibrio).	Spiral - shaped rods; ærobic.	. . . . . .

BACTERIA.—Continued.

Culture Characters.	Actions.	Habitat.	Discoverer.
Not cultivated.	Supposed to cause syphilis.	In tissue and secretions of syphilities.	Lustgarten.
Liquefy gelatin slowly; colonies have radiated appearance; a thorny growth along the track in test-tube.	Produces tetanus in man and animals.	Earth and manure.	Nicolaier and Kitasato.
Not liquefying; little porcelain-like disks; thick slimy layer on potato.	Fatal to guinea pigs and white mice.	Found in cavernous phthisical lungs.	Gaffky.
. . . .	Supposed to be the cause of *Rhus* (poison ivy) poisoning.	Found in the *Rhus toxicodendron*.	Burrill.
Along needle-line, white, wreath-like arrangement of small spheres, turning yellow; best at 37° C.; not liquefying.	In rabbits no result, but on human cornea typical trachoma.	Found in follicles of Egyptian eye disease.	Sattler and Michel.
Grows best on blood-serum and glycerin agar at 37° C., forming little white crumbs on the surface; under microscope a hairy matted coil is seen; growths on potatoes when air-tight have been obtained.	Causes tuberculosis, local and general, in man and lower animals.	In all organs and secretions of tubercular persons.	Koch.
. . . .	. . . . . .	In caseous nodules of tubercular animals.	Malassez and Vignal.
Not liquefying; little whetstone-shaped yellow colonies in the deep, and leaf-shaped ones on the surface; on potato, a very transparent, moist layer.	Gives rise to enteric or typhoid fever in man.	Found in dejecta and spleen and urine of typhoid patients.	Eberth.
Liquefy rapidly; small round colonies; dark funnel-shaped liquefaction in test-tube.	Several animals have died from inoculations.	From old cheese.	Dencke.

PLATE III.

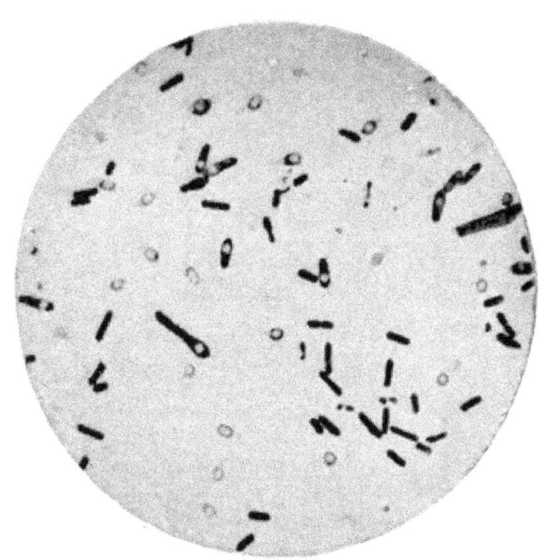

BACILLI OF SYMPTOMATIC ANTHRAX, WITH SPORES 1000 X.
(Fränkel and Pfeiffer.)

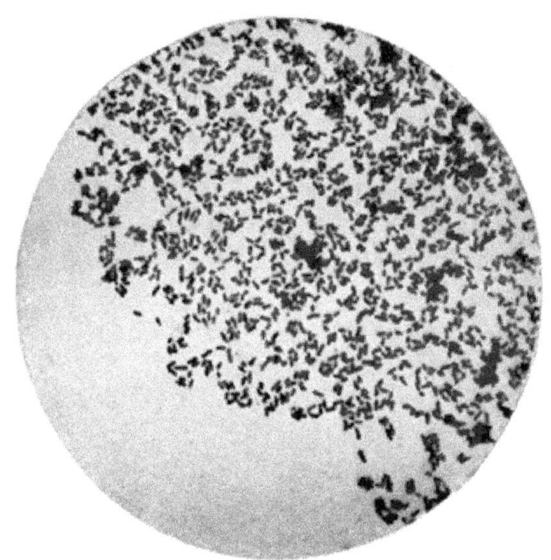

DIPHTHERIA BACILLUS PURE CULTURE 1000 X.
(Fränkel and Pfeiffer.)

PLATE IV.

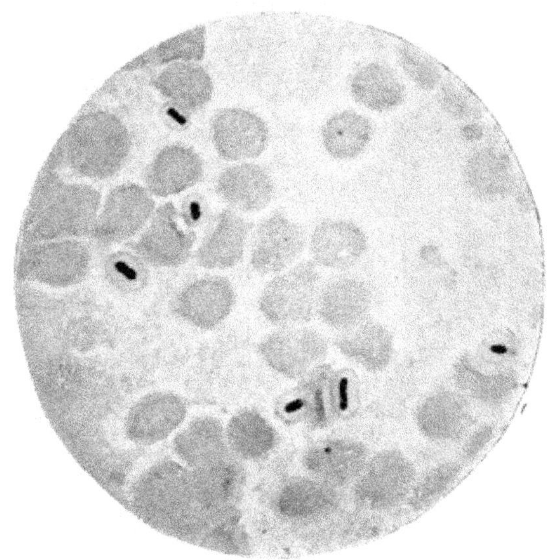

PFEIFFER'S CAPSULE BACILLUS IN BLOOD 1000 X.
(Fränkel and Pfeiffer.)

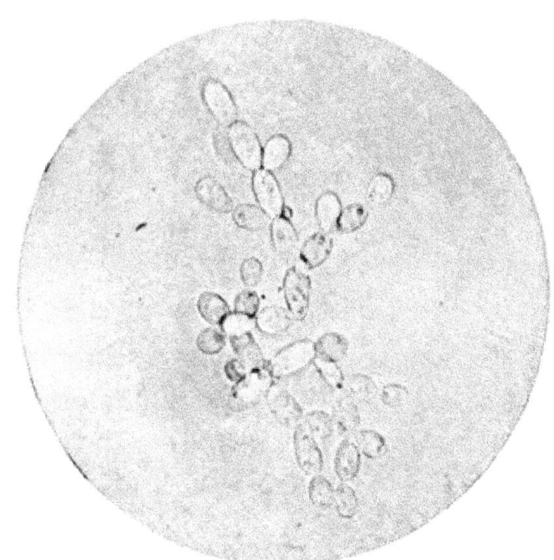

YEAST-CELLS 500 X.
(Fränkel and Pfeiffer.)

PLATE V.

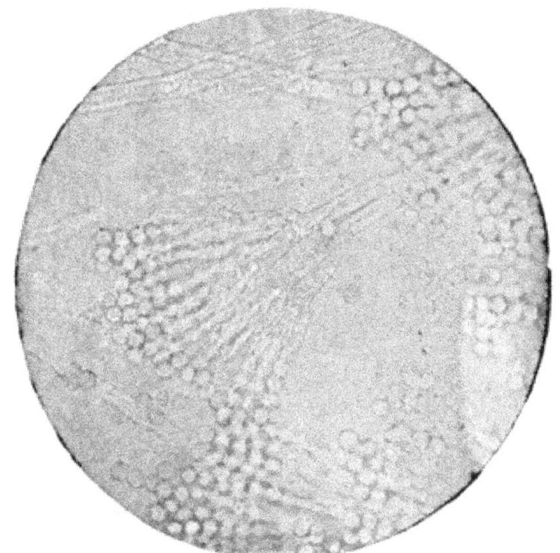

PENICILLIUM GLAUCUM 500 X.
(Fränkel and Pfeiffer.)

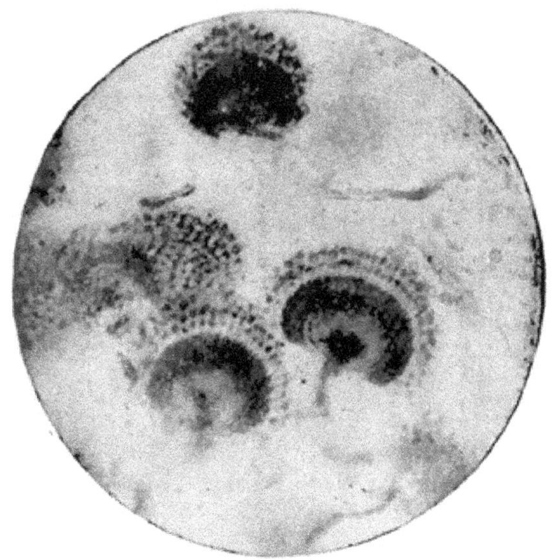

ASPERGILLUS FUMIGATUS 500 X.
(Fränkel and Pfeiffer.)

INDEX.

www.ingramcontent.com/pod-product-compliance
Lightning Source LLC
Chambersburg PA
CBHW021706210326
41599CB00013B/1538

* 9 7 8 3 3 3 7 1 7 6 9 7 6 *